高等学校“十二五”规划教材

有机化学实验

Experimental Organic Chemistry

（英汉双语教材）

陈 彪 魏永慧 编

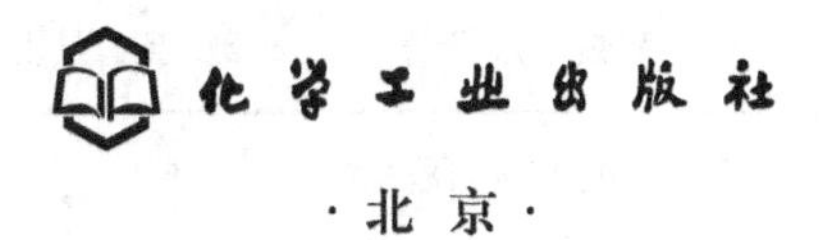

·北京·

本书是编者在总结多年来有机化学实验的教学经验及近年来实施双语教学实践的基础上，参考了国内外出版的同类教材，采用英汉两种语言，编写的与有机化学双语课程配套使用的《有机化学实验》教材。

全书包括：有机化学实验的一般知识，有机化学实验的基础知识和基本操作，有机化学综合实验（包括合成及分离提取两大部分），附录及参考文献。

本书可使学生在掌握有机化学实验基本原理、基本技能的同时，提高英语实际应用能力和水平，适用于化学、医学、药学及相关专业的学生使用，也可作为有机化学及相关专业硕士研究生的实验参考书。

图书在版编目（CIP）数据

有机化学实验（英汉双语教材）/陈彪，魏永慧编．北京：化学工业出版社，2013.7（2022.1重印）
高等学校“十二五”规划教材
ISBN 978-7-122-17314-0

Ⅰ.①有… Ⅱ.①陈…②魏… Ⅲ.①有机化学-化学实验-高等学校-教材-汉、英 Ⅳ.①O62-33

中国版本图书馆CIP数据核字（2013）第097054号

责任编辑：宋林青　　文字编辑：向　东
责任校对：边　涛　　装帧设计：史利平

出版发行：化学工业出版社（北京市东城区青年湖南街13号　邮政编码100011）
印　　装：北京虎彩文化传播有限公司
787mm×1092mm　1/16　印张13¾　字数339千字　2022年1月北京第1版第4次印刷

购书咨询：010-64518888　售后服务：010-64518899
网　　址：http://www.cip.com.cn
凡购买本书，如有缺损质量问题，本社销售中心负责调换。

定　　价：36.00元

前　言

为了适应高等教育事业的快速发展，满足21世纪高等教育教学改革发展的需要，全面落实教育部提出的本科教学质量工程，进一步提高学生动手能力、实践技能及英语水平，提高教育教学和人才培养质量及人才的国际竞争能力，我们在总结多年来有机化学实验的教学经验及近年来实施双语教学实践的基础上，参考了国内外出版的同类教材，注重实验的小量化、绿色化，采用以英文为主、中英文相对照的方式，编写了这本与有机化学双语课程配套使用的《有机化学实验》英汉双语教材。

本书在编写过程中努力体现“反映特色，加强基础，注意交叉，够用为度，注重能力培养”的现代课程建设理念。

书中对基本操作和实验方法用英语作了较为详细精炼的描述。为了加强基本实验技能的训练，加深学生对实验原理和实验操作的理解，在有关章节中均附有较为详细的注释、思考题，以便于教学或学习。本书中多数的合成、提取与分离实验是编者多年来教学研究、科学研究与实践所形成的较成熟的实验，教学效果良好，对某些毒性较大、内容陈旧的传统实验项目进行了改革和更新。

在内容和结构安排上，从有机化学实验的基本知识入手，采取循序渐进的方式，主要包括有机化学实验基础知识、基本操作、综合实验三大部分。教材中的综合实验可加深学生对本门课程教学内容的全面了解和掌握，有利于学生综合素质和创新能力的培养。

本书在编写过程中参考了部分国内外出版的同类教材，在此深表感谢。

由于编者水平有限，书中难免有疏漏、欠妥之处，敬请各位专家、老师和读者批评赐教。

编者

2013年2月

前言

目 录

Chapter 1 Introduction of Experimental Organic Chemistry

Experimental organic chemistry is an integral and basic part of the organic chemistry course. With the coming of new techniques, this course is being directed towards the development of small-scale experiments, high-efficient operations and the use of environment-friendly chemicals. The purpose of this course is to provide an opportunity to observe the reality of compounds and reactions, learn something of the operations and techniques that are used in experimental organic chemistry and in other areas, and further understand the basic principles of organic chemistry. Students should get into the habit of "preparation (pre-lab)-experiment-recording (in-lab)—summary (post-lab)", and rigorous scientific attitude, and work style.

1.1 Basic Rules for the Organic Chemistry Lab

In order to ensure all experiments go smoothly and laboratory safety is observed, students must abide by the following rules when entering into an organic lab:

① Familiarize yourself with the safety rules for lab work and learn about how to correctly use water, power, gas, hood, fire extinguisher and so on. Get to know what to do in the event of experimental accidents. Everyone, before doing the experiment, should be well prepared, understand the hazardous nature and safe usage of chemicals and promote safety consciousness. The experimental instruments and equipment must be used with care, adhering to their operating procedures. Report all abnormal conditions to your instructor to minimize the operational hazards.

② Before doing an experiment, check all glass equipment. During the experiment, use it carefully and skillfully; after the experiment, clean it up and keep it in order.

③ In the experiment, keep your experimental area and the whole lab tidy, operate with care, and adhere to the experimental procedures as well as reagent specifications and dosage required in every experiment. If you want to make any change, ask your instructor to get authorization. Never leave an ongoing experiment unattended.

④ Before using chemicals, read their labels carefully. Use them only as required in the experiment. Cover the stopper of the container immediately after use, and avoid the stoppers being confused as well as the chemicals being contaminated. Don't leave a mess for someone else to clean up. Don't change the position at random of normal reagents and common instruments in the lab such as balances, desiccator, refractometer and so on.

⑤ Your full attention must be given to what you are doing during the experimental period. Don't be careless or clown around in lab. You may hurt yourself or other people. Don't speak loudly or eat or drink in the lab.

⑥ In-lab or post-lab, all kinds of solid or liquid waste should be placed in various authorized containers.

⑦ Before leaving lab, check carefully whether water, power and the gas are switched off safely, and wash your hands thoroughly with soap and water.

1. 2 Common Lab Equipment and Apparatus

1. 2. 1 Lab Equipment

A typical set of lab equipment including glassware with standard-taper ground glass joints and non-glass equipment is shown in Figure 1. 1～Figure 1. 4.

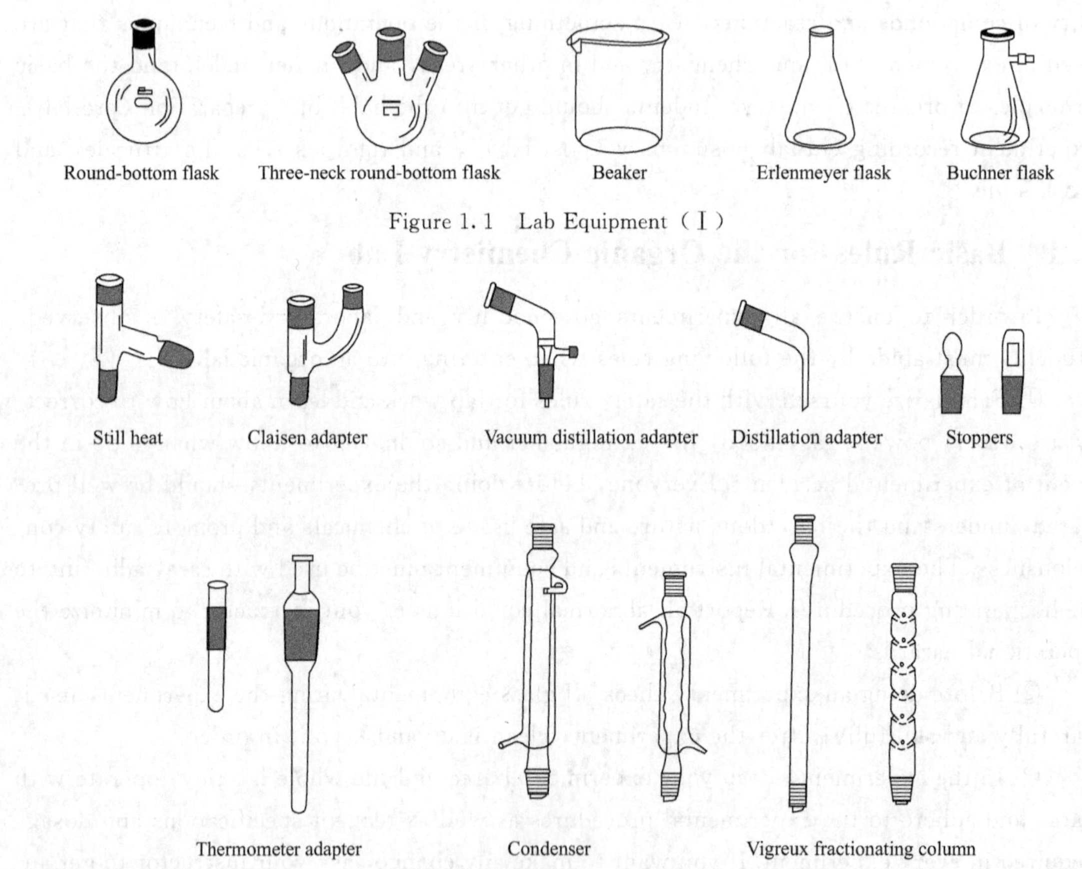

Figure 1. 1 Lab Equipment（Ⅰ）

Figure 1. 2 Lab Equipment（Ⅱ）

Round-bottom flask is used for distillation, reflux; three-neck round-bottom flask is used for more complicated reaction set-ups (two-neck flask are also available); Beaker is used for heating or mixing; Flask is used for titration, crystallization, preparation; Buchner flask is used for collecting the filtrate.

Still heat is used for distillation; Various adapter is used for distillation, vacuum distillation; Condenser is used for distillation; Air condenser is used for distillation with high boiling liquids; Vigreux fractionating column is used for fractional distillation.

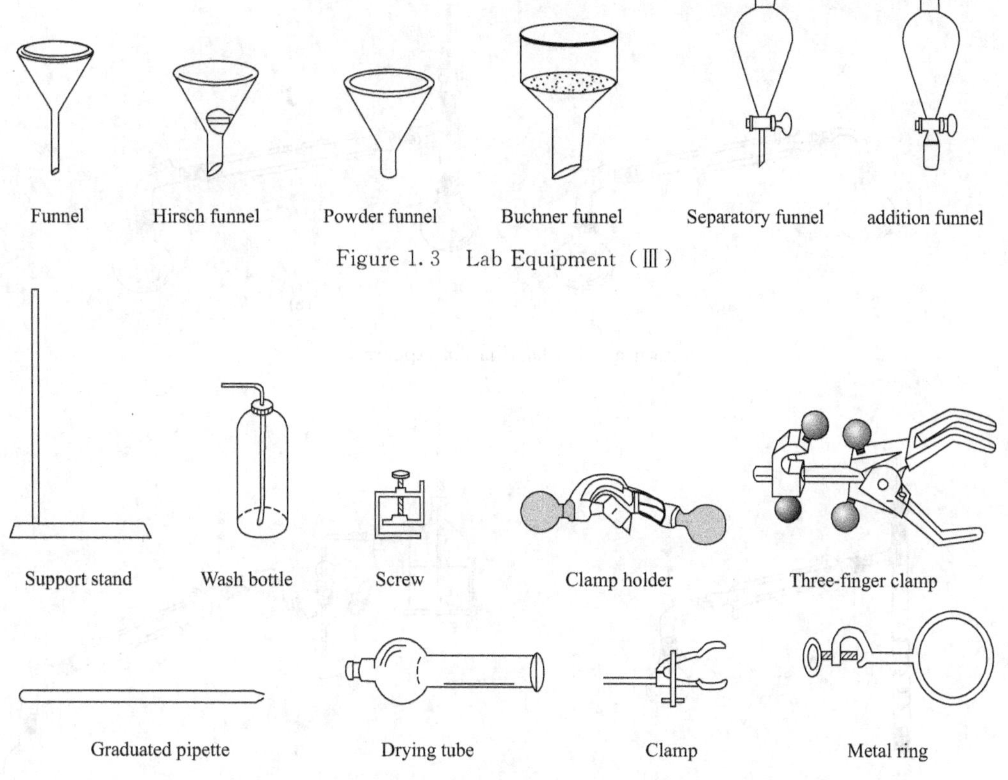

Figure 1.3 Lab Equipment (Ⅲ)

Figure 1.4 Lab Equipment (Ⅳ)

Various funnel is used for different filtration; Addition funnel is used for adding liquids; Separatory funnel is used for extraction and reaction.

Drying tube is used for drying gases. Support stand, clamp holder and clamp are all used for fixing.

1.2.2 Common Apparatus

All kinds of common apparatus are shown in Figure 1.5～Figure 1.11.

The rotary evaporator is used for the removal of volatile solvent from solution, leaving behind the non-volatile component (see Figure 1.10).

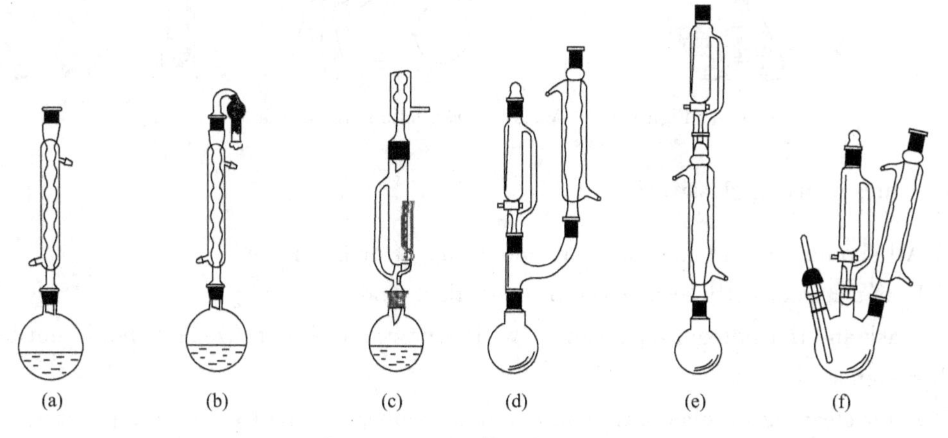

Figure 1.5 Reflux Apparatus

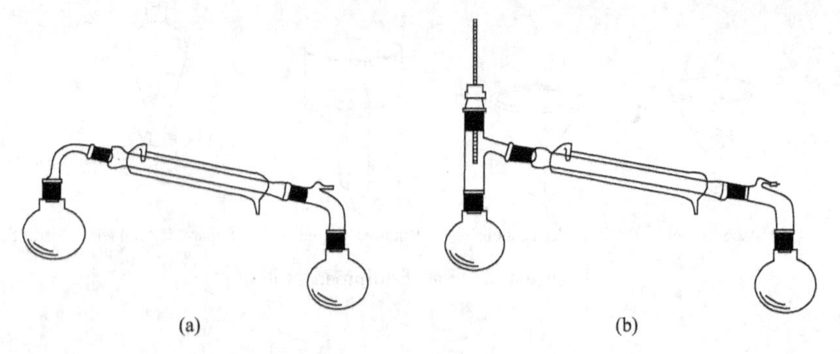

Figure 1.6 Distillation Apparatus

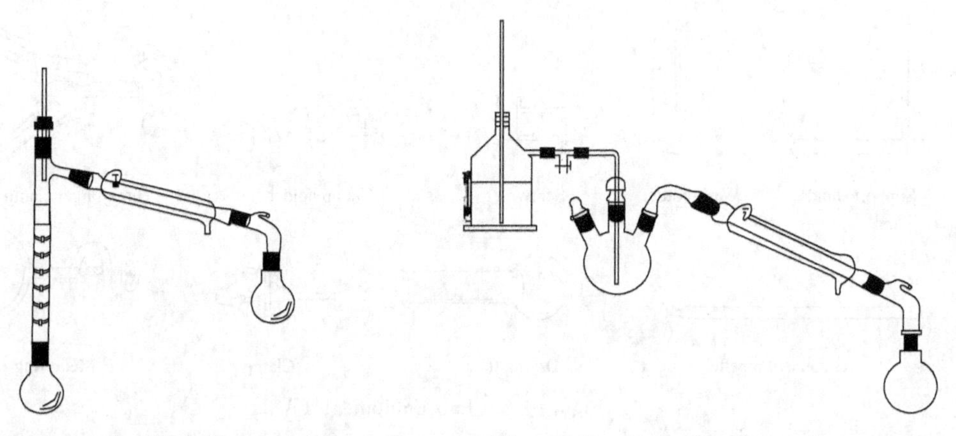

Figure 1.7 Fractional Distillation Apparatus

Figure 1.8 Steam Distillation Apparatus

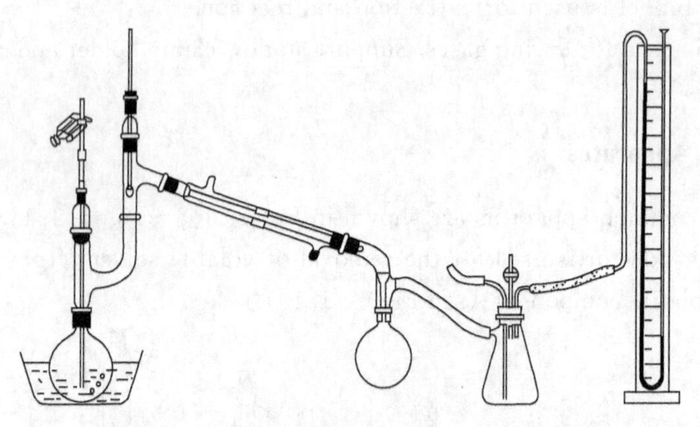

Figure 1.9 Vacuum Distillation Apparatus

1.2.3 Notes of using glassware

① All should be used carefully, avoiding impact or breakage.

② Don't heat directly except the beaker, flask and tube.

③ Flask and flat-bottom flask cannot withstand reduced pressure and should not be used in such systems.

④ After cleaning up glassware containing a stopper, a small piece of paper must be put between the stopper and ground joint to avoid adhesion.

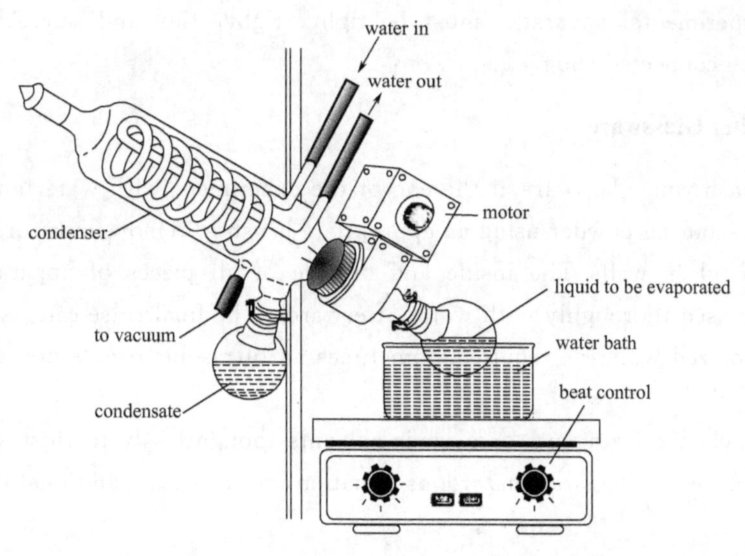

Figure 1. 10 A Rotary Evaporator with Condenser and Receiving Flask

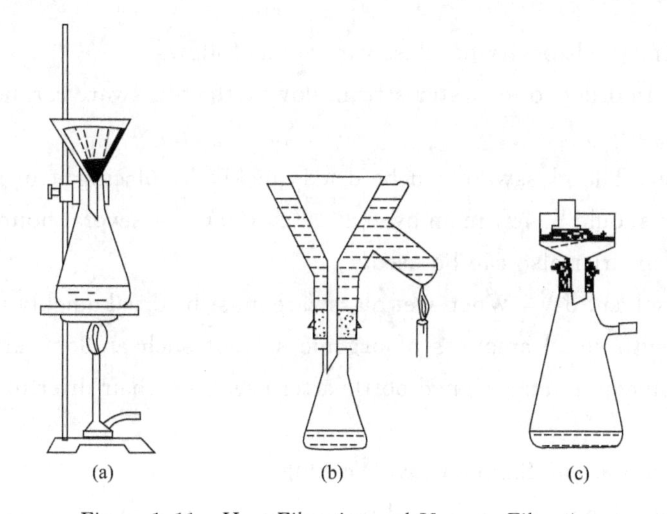

Figure 1. 11 Heat Filtration and Vacuum Filtration

⑤ The glass of a mercury bulb is thin and ease-to-break, thus should be used with care. Never use it as a stirring rod. After use, cool it down, and rinse it afterwards to keep it from cracking. The measurement of the thermometer doesn't go beyond its graduated range.

1. 2. 4 Notes of assembling apparatus

① All glassware and accessories must be clean and fitted properly.

② When assembling the apparatus follow the principle of "bottom-to-top, left-to-right", step by step.

③ When disassembling, observe the rule of "right-to-left, up-to-down", one by one.

④ A reaction apparatus under the ordinary pressure must have an opening to the atmosphere to avoid development of a dangerously high pressure within the system when heat is applied.

⑤ All experimental apparatus must be tight, right, tidy and safe. All ground-glass joints should be connected snugly.

1.2.5 Cleaning Glassware

Always wash your glassware at the end of the experiment with water and either detergent or a mild scouring powder using an appropriate brush to remove most organic chemicals adhered to the glass walls. The inside and outside of all pieces of apparatus should be scrubbed and rinsed thoroughly with water afterwards. The final rinse can also be done with distilled or deionized water as required. Sometimes an ultrasonic oscillators might be useful for cleaning.

Never use chemical reagents or organic solvents thoughtlessly to rinse glassware. This may produce waste, and create a hazardous situation, resulting in additional pollution to our environment.

1.2.6 Drying Glassware

The common methods of drying glassware are as follows:

(1) Air dry In order to let water stream down, the glassware can be left upside down on a drying rack to dry.

(2) Oven dry The glassware can be dried quickly by placing it in an oven. For complete drying, glass should be left in an oven at 110～120℃ for several hours. Besides this, an air flow drier or hair drier also can be used.

(3) Organic solvent dry When wet glassware must be dried quickly for immediate use, it may be rinsed with small amounts of organic solvent such as 95% ethanol or acetone, which must be drained into an assigned bottle after use. Use a hair drier to evaporate the solvent afterwards.

Experiment: Practice on Simple Glass Working

Cutting and bending a glass tube and rod as well as preparing a capillary are all the primary operations in organic experiment.

(1) Cutting a Glass Tube and Rod

① Place the glass tube (or rod) flat on the table. Hold the tube firmly and create a scratch on the glass wall surface by drawing a tiny sand wheel or three corner file perpendicularly across it a couple of times (Do not saw back and forth. A short, single sharp scratch is more likely to produce a clean even edge. At the same time, keep the tube dry).

② Hold the tube in both hands, one on each side of the scratch. Keep your thumbs as close as possible to the scratch away from your body. Apply gentle pressure with your thumbs behind the scratch and push with the fingers in the direction of its length. The glass tube should break cleanly at the scratch (Keep them away from your eyes when broken).

③ Light the burner and get a hot blue flame. Pass one end of tube or rod through the flame a few times, rotating the tube until the rough edges become smooth[1].

(2) Bending Glass Tube

① Light the burner and open the air hole. This gives a hot blue flame. Using both hands, rotate the dry tube in a declining position back and forth through the top of the flame.

② Keep rotating the tube until it glows red and has become soft. Then take the tube out of the flame and bend to the desired angle.

③ Don't twist the tube while heating and bending in order to ensure that the two ends of the bent tube remain in the same plane.

④ It is necessary to get a smooth bend without any kinks or constrictions. Check whether the bend is at the required angle (e. g. 90°) and whether the whole tube is in one plane[2] after the tube is bent. Then put it on the asbestos mat to cool (Don't put it on the table directly).

(3) Procedure

① Take several used tubes to practise the basic operation of cutting, heating, rotating and bending, to improve your glassworking skills.

② Preparation of a stirring rod: take a glass rod of 5mm diameter and 20cm length. Heat the two ends of the tube respectively in a hot blue flame until each end of it becomes round shaped.

③ Bending the tube: take two tubes of 5mm diameter, and bend them into 90°and 70° tubes respectively[3].

Notes

[1] Do not heat too much and do not stop rotating. Place the hot glass on an asbestos pad to cool. When the tube is cool, fire polish the other cut end. Again allow to cool. Be careful of the rough edges of the tube which sometimes may cut your hands

[2] After bending the tube, anneal it in a small flame and then put it on asbestos pad directly

[3] The length of the glass tube is changeable as required.

1.3 General Lab Safety

Generally speaking, organic experiments utilize mainly glassware, chemicals and electrical appliances, all of which can do harm to the human body and the environment if used improperly. Chemicals are hazardous because of their flammable, explosive, volatile, corrosive and toxic properties. Also, there is the possibility of experimental accidents to glass equipment and electrical appliances if operated incorrectly. Therefore, organic lab is potentially one of the most dangerous locations for students.

1.3.1 Fire-proof

The experimental operation must be normalized and the apparatus must be assembled correctly. Flammable, explosive and volatile chemicals mustn't be discarded randomly and must be recovered specifically after the experiment. They should be kept away from an open flame. In case of a fire, first of all, cut power and the gas off, move the flammable and explosive reagents away, and then put the fire out in a proper way using a fire extinguisher, asbestos cloth, covering with sand, or rushing water and so on.

1.3.2 Explosion-proof

The apparatus should be assembled correctly. The whole system should not be made tight in the process of normal distillation and reflux. Distillation to dryness is also a dangerous practice because of the possible presence of peroxides or other explosive materials in the dry residue in the flask. The glassware and apparatus should be checked first to determine whether it can withstand the system pressure before vacuum distillation. If you don't add any boiling chips when starting distillation, stop heating immediately and re-add them after cooling. Keep the cooling water moving smoothly during distillation.

A fierce explosion or combustion can be produced when some organic compounds come into contact with oxidizers. Beware of their handling and storage.

1.3.3 Poisoning-proof

There are different ratings of toxicity among most organic reagents. The experiment with an irritant or toxic gas discharged must be always carried out in a hood or in a well-ventilated circumstance, or using a gas trap.

The manipulation of toxic or corrosive chemicals should follow the designated procedures strictly. Don't touch or come into contact directly with them. Keep them away from your mouth or cuts or abrasions of the skin, and never pour them into the sewer.

If you have some poisoning symptoms such as dizziness, headache, or other symptoms during the experiment you should leave the laboratory area and move to an area where you can breath fresh air and rest. In case of the poisoning is severe or symptoms persist, you should receive medical treatment.

1.3.4 Prevent Chemical Burns

Chemicals such as strong acids, strong bases, bromine, etc, should be used with great care in order to avoid contact with your skin which could cause chemical burns. In case of such an accident, wash the affected area immediately with copious amounts of running water, and then further treatments as follows.

Acid-injury: use 1% $NaHCO_3$ solution for the eye-wash and 5% $NaHCO_3$ solution for skin-wash.

Base-injury: use 1% boric acid for the eye-wash and 1%～2% acetic acid for skin-wash.

Bromine-injury: wash immediately with alcohol, and smear with glycerol or coat with a scald ointment.

If the situation is severe, go to hospital after first aid.

1.3.5 Cuts and Scalds

An accident involving cut or scald occurs in the use of glassware or manipulation of glassware if operated improperly. In case of such an accident, deal with it by the following methods.

Cuts: Cuts from broken glass are a constant potential hazard during experiments. The cut should be rinsed thoroughly with running water or hydrogen peroxide for a while to ensure that all tiny pieces of glass are removed. After this, wipe the cut with merbromin and bind up with gauze; if the cut is severe, first bind up with gauze and then send the patient to the hospital.

Scalds: Smear some scald ointment on the affected area if the situation is just a bit superficial; coat with scald ointment and go to the hospital for further treatment if the situation is severe.

1.4 Disposal of Lab Waste

Experimental operations always generate different kinds of solid or liquid waste. Waste disposal has been one of the major environmental problems of modem society. Special measures should be taken to observe national regulations and local organic lab rules of waste disposal. The handling of such wastes in the lab can be done in the following way:

① All waste generated in the lab can be classified into solid or liquid waste, and hazardous or nonhazardous waste, and disposed of properly. Some hard-to-handle hazardous waste should be delivered to the environmental department for special treatment.

② Small amounts of acids such as hydrochloric, sulfuric, and nitric, or bases such as sodium or potassium hydroxide, should be neutralized first and diluted with large amounts of water before flushing down the drain.

③ Organic solvents should be poured into properly labeled waste containers and stored in a well-ventilated place.

④ Nonhazardous solid waste such as paper, broken glass, corks, alumina, silica gel, magnesium sulfate, calcium chloride, and so on, should not be blended with other hazardous waste, and can probably go into the ordinary dustbin. Hazardous solid waste should be disposed of in a labeled container. The exact name of the contents should be written on the label.

⑤ Chemicals that react violently with water should be decomposed in a suitable way in a hood before disposal.

⑥ Some carcinogens and substances suspected of causing cancer must handled with great care, avoiding contact with your body.

1.5 Common Organic Solvents

(1) Ethyl Ether

Additional name(s): Ethoxyethane; ether; diethyl ether; ethyl oxide; diethyl oxide. Molecular formula: $C_4H_{10}O$. Molecular weight: 74.12. Elemental analysis: C 64.82%, H 13.60%, O 21.59%. Line Formula: $C_2H_5OC_2H_5$. Properties: Mobile, very volatile, highly flammable liquid; Explosive! Characteristic, sweetish, pungent odor; Burning taste. Tends to form explosive peroxide under the influence of air and light, esp. when evaporation to dryness is attempted. Peroxides may be removed from ether by shaking with 5% aq ferrous sulfate solution. Addition of naphthols, polyphenols, aromatic amines, and aminophenols has been proposed for the stabilization of ethyl ether. d_4^{20} 0.7134; m.p. −116.3℃ (stable crystals); b.p. 34.5℃; Flash point, closed cup: −49℃. Ether is slightly sol in water and water is slightly sol in ether. Miscible with lower aliphatic alcohols, benzene, chloroform, other fat solvents, many oils. Caution: Keep away from fire; Potential symptoms of overexposure are dizziness; drowsiness; headache, excitedness and narcosis; nausea, vomiting; irritation of eyes, upper respiratory system and skin.

(2) Ethyl Alcohol

Additional name(s): Ethanol; absolute alcohol; anhydrous alcohol; dehydrated alcohol; ethyl hydrate; alcohol. Molecular formula: C_2H_6O. Molecular weight: 46.07. Elemental analysis: C 52.14%, H 13.13%, O 34.73%. Line Formula: CH_3CH_2OH. Properties: Clear, colorless, very mobile, flammable liquid; pleasant odor; burning taste; absorbs water rapidly from air. d_4^{20} 0.789; b.p. 78.5℃; m.p. −114.1℃; Flash point, closed cup: 13℃. n_D^{20} 1.361. Miscible with water and with many organic liquids. Caution: Keep tightly closed, cool, and away from flame!

(3) Benzene

Molecular formula: C_6H_6. Molecular weight: 78.11. Elemental analysis: C 92.26%, H 7.74%. Properties: Clear, colorless, volatile, highly flammable liquid; characteristic odor. d_4^{15} 0.8787; b.p. 80.1℃; m.p. 5.5℃; n_D^{20} 1.5011; Flash point, closed cup: −11℃. Slightly sol in water. miscible with alcohol, chloroform, ether, acetone, glacial acetic acid, carbon disulfide, oils. Caution: Keep in well-closed containers in a cool place and away from fire. Potential symptoms of overexposure by inhalation or ingestion are dizziness, headache, vomiting, visual disturbances, staggering gait, hilarity, fatigue, CNS depression, and loss of consciousness, respiratory arrest. Chronic exposure has been associated with bone marrow depression and leukemia. Direct contact may cause irritation of eyes, nose, respiratory system and skin; dermititis may develop due to defatting action. Aspiration into the lung may lead to chemical pneumonitis.

(4) Toluene

Addition name(s): Methylbenzene; phenylmethane. Molecular formula: C_7H_8. Molecular weight: 92.14. Elemental analysis: C 91.25%, H 8.75%. Properties: Flammable, refractive liq; benzene-like odor. d_4^{20} 0.866; m.p. −95℃; b.p. 110.6℃; n_D^{20} 1.4967; Flash point, closed cup: 4.4℃, Very slightly sol in water; miscible with alcohol, chloroform, ether, acetone, glacial acetic acid, carbon disulfide. Caution: Readily absorbed by inhalation ingestion and somewhat by skin contact. Direct contact may cause severe dermatitis due to drying and defatting action. May present lung aspiration hazard if ingested. Potential symptoms of acute overexposure by inhalation may include local initiation; CNS excitation and depression. Low concentrations may result in transitory mild upper respiratory tract initiation, mild eye irritation, lacrimation, metallic taste, slight nausea, hilarity, lassitude, drowsiness and impaired balance. High concentrations may cause paresthesia, vision disturbances, dizziness, nausea, headache, narcosis and collapse; death from respiratory failure or sudden ventricular fibrillation. Chronic overexposure by inhalation has been associated with hepatotoxicity and nephrotoxicity. Syndromes following chronic inhalation involve severe muscle weakness, cardiac arrhythmias, gastrointestinal and neuropsychiatric complaints.

(5) Acetone

Additional name(s): 2-Propanone; dimethylketone; beta-keto propane; pyroacetic. Molecular fomula: C_3H_6O. M olecular weight: 58.08. Elemental analysis: C 62.04%, H 10.41%, O 27.55%. Line Formula: CH_3COCH_3. Properties: Volatile, highly flammable liquid; characteristic odor; pungent, sweetish taste. d_{25}^{25} 0.788; b.p. 56.5℃; m.p. −94℃; n_D^{20} 1.3591. Flash point, closed cup: −18℃. Miscible with water, alcohol, dimethylformamide, chloroform, ether, most oils. Caution: Keep away from fire! Keep away from plastic eyeglass frames, jewelry, pens and pencils, rayon stockings and other rayon garments.

(6) Methanol

Additional name(s): Methyl alcohol; carbinol. Molecular formula: CH_4O. Molecular weight: 32.04. Elemental analysis: C 37.48%, H 12.58%, O 49.93%. Line Formula: CH_3OH. Properties: Flammable, poisonous, mobile liq. Slight alcoholic odor when pure; crude material may have a repulsive, pungent odor. Burns with anon-luminous, bluish flame. d_4^{20} 0.7915; m.p. −97.81℃. b.p. 64.7℃; n_D^{20} 1.3292; Flash point, closed cup: 12℃. Miscible with water, ethanol, ether, benzene, ketones and most other organic solvents. Caution: Poisoning may occur from ingestion, inhalation or percutaneous absorption. Acute Effects: Headache, fatigue, nausea, visual impairment or complete blindness (may be permanent), acidosis, convulsions, mydriasis, circulatory collapse, respiratory failure, death.

(7) Ethyl Acetate

Additional name(s): Acetic acid ethyl ester; acetic ether. Molecular formula: $C_4H_8O_2$. Molecular weight: 88.11. Elemental analysis: C 54.53%, H 9.15% O 36.32%. Line Formula: $CH_3COOC_2H_5$. Properties: Clear, volatile, flammable liq; characteristic fruity odor; pleasant taste when diluted. Slowly dec by moisture, then acquires an acid reaction. Absorbs water (up to 3.3% w/w). d_4^{20} 0.902; d_{25}^{25} 0.898; b.p. 77℃; m.p. −83℃; Flash point 7.2℃ (open cup). Explosive limits (% vol in air): 2.2 to 11.5. n_D^{20} 1.3719. 1mL dissolves in 10mL water at 25℃; Miscible with alc, acetone, chloroform, ether. Caution: Keep tightly closed in a cool place and away from fire. Potential symptoms of overexposure are irritation of eyes, nose and throat, narcosis, dermatitis.

(8) Chloroform

Additional name(s): Trichloromethane. Molecular formula: $CHCl_3$. Molecular weight: 119.38. Elemental analysis: C 10.06%, H 0.845% Cl 89.09%. Properties: Highly refractive, nonflammable, heavy, very volatile, sweet-tasting liquid; characteristic odor. d_{20}^{20}

1.484; b.p. 61～62℃; m.p. −63.5℃; n_D^{20} 1.4476. Slightly sol in water; Miscible with alcohol, benzene, ether, petroleum ether, carbon tetrachloride. Caution: Potential symptoms of overexposure are dizziness, mental dullness, nausea and disorientation; headache, fatigue; anesthesia; hepatomegaly; direct contact may cause irritation of eyes and skin.

(9) Methylene Chloride

Additional name(s): Dichloromethane; methylene dichloride. Molecular formula: CH_2Cl_2. Molecular weight: 84.93. Elemental analysis: C 14.14%, H 2.37%, Cl 83.48%. Properties: Colorless liquid; vapor is not flammable and when mixed with air is not explosive. Soluble in approximately 50parts water; miscible with alc, ether, DMF; b.p. 39.8℃; m.p. −95; d_4^{20} 1.3255; n_D^{20} 1.4244. Caution: Potential symptoms of overexposure are fatigue, weakness, sleepiness, lightheadedness; numbness or tingle of limbs; nausea; irritation of eyes and skin.

(10) *n*-Hexane

Molecular formula: C_6H_{14}. Molecular weight: 86.18. Elemental analysis: C 83.63%, H 16.17%. Line Formula: $CH_3(CH_2)_4CH_3$. Properties: Colorless, very volatile liquid; faint, peculiar odor. d_{20}^{20} 0.660; b.p. 69℃; m.p. −100～−95℃; n_D^{20} 1.375. Insol in water; miscible with alcohol, chloroform, ether. Caution: Potential symptoms of overexposure are light-headedness; nausea, headache; numbness of extremities, muscle weakness; irritation of eyes and nose; dermatitis; chemical pneumonia; giddiness.

(11) Cyclohexane

Additional name(s): Hexamethylene. Molecular formula: C_6H_{12}. Molecular weight: 84.16. Elemental analysis: C 85.16%, H 14.37%. Properties: Flammable liq. Solvent odor. Pungent when impure. d_4^{20} 0.7781; b.p. 80.7℃; m.p. 6.47℃; n_D^{20} 1.4264. Flash point, closed cup: −18℃, Flammability limits in air 1.3%～8.4% *v/v*. Insol in water; miscible with ethanol, ethyl ether, acetone, benzene, carbon tetrachloride. Caution: Potential symptoms of overexposure are irritation of eyes and respiratory system; drowsiness; dermatitis; narcosis; coma.

(12) Tetrahydrofuran

Additional name(s): Diethylene oxide; tetramethylene. Molecular formula: C_4H_8O. Molecular weight: 72.11. Elemental analysis: C 66.63%, H 11.18%, O 22.19%. Properties: Liquid; Ether-like odor. d_4^{20} 0.8892; m.p. −108.5℃; b.p. 66℃; n_D^{20} 1.4070. Miscible with water, alcohols, ketones, esters, ethers, and hydrocarbons. Caution: Distil only in presence of a reducing agent, such as ferrous sulfate; peroxide explosions have occurred. Potential symptoms of overexposure are irritation of eyes and upper respiratory system; nausea, dizziness and headache; may cause skin irritation.

(13) *N*,*N*-Dimethylformamide

Additional name(s): DMF; Molecular formula: C_3H_7NO. Molecular weight: 73.09. Elemental analysis: C 49.30%, H 9.65%, N 19.16%, O 21.89%. Line Formula: HCON $(CH_3)_2$. Properties: Colorless to very slightly yellow liquid; Faint amine odor; m.p. −61℃; b.p. 153℃; d_4^{25} 0.9445; n_D^{25} 1.42803; Flash point, open cup: 67℃. Miscible with water and most common organic solvents. Caution: Potential symptoms of overexposure are nausea, vomiting and colic; liver damage, hepatomegaly; high blood pressure; facial flush; dermatitis.

(14) Acetonitrile

Additional name(s): Methyl cyanide; cyanomethane. Molecular formula: C_2H_3N. Molecular weight: 41.05. Elemental analysis: C 58.52%, H 7.37%, N 34.12%. Line Formula: CH_3CN. Properties: Liquid; Ether-like odor. Poisonous! m.p. −45℃; b.p. 81.6℃; Flash point 12.8℃; d_4^{15} 0.78745; n_D^{15} 1.34604. Slightly sol in water; miscible in methanol, methyl acetate, ethyl acetate, acetone, ether, acetamide solutions, chloroform, carbon tetrachloride, ethylene chloride and many unsaturated hydrocarbons. Caution: Potential symptoms of overexposure are asphyxia; nausea, vomiting; chest pain; weakness; stupor, convulsions. Direct contact may cause skin and eye irritation.

(15) Pyridine.

Molecular formula: C_5H_5N. Molecular weight: 79.10. Elemental analysis: C 75.92%, H 6.37%, N 17.71%. Properties: Flammable, colorless liq; characteristic disagreeable odor; sharp taste. d_4^{20} 0.98272; m.p. −41.6℃; b.p. 115.2℃; Flash point, closed cup: 20℃; n_D^{20} 1.50920. Miscible with water, alcohol, ether, petroleum ether, oils and many other organic liquids. Good solvent for many organic and inorganic compounds. Weak base; forms salts with strong acids. Caution: Potential symptoms of overexposure are headache, nervousness, dizziness and insomnia; nausea, anorexia; frequent urination; eye irritation; dermatitis; liver and kidney damage.

1.6 Experimental Preview Record and Laboratory Report

(1) Experimental Preview Record

Make good preparations before you come to the lab by reviewing the theory and seeking information about the chemicals involved. If you prepare well, you will save time and effort. In general, the experimental record includes the following parts:

① Experimental title and general statement of the experimental apparatus and process to be done.

② Note any special observations or precautions required.

③ Jot down any chemical or biological hazards it presents.

(2) Experimental Report

Usually, the experimental report should cover the following parts:

① Experiment and title.

② Purpose of the experiment: main techniques introduced by the experiment.

③ Principle of the experiment: chemical reaction; balanced chemical equations or reaction mechanism.

④ Experimental apparatus and reagent.

⑤ Experimental procedure: process and experimental observations record

⑥ Result and discussion: analysis of your data and brief comment on sources of error in your experiment.

⑦ Conclusions.

⑧ Questions.

1.7 Chemical Literature

The abundance of organic chemistry literature can be overwhelming. With the wide spread use of computers, the search of organic chemistry literature is getting easier and faster. In this part, selected examples from different literature sources are given, along with the brief remarks.

(1) Laboratory Handbooks

① CRC Handbook of Chemistry and Physics (annual editions), published by the CRC Press, Boca Raton, FL. It contains about 15, 000 organic compounds with physical properties. Organic compounds are alphabetically ordered. This handbook is cross-referenced to Beilstein Handbook and contain a molecular formula index.

② Merck Index of Chemicals and Drugs (12th). An Encyclopedia of Chemicals, Drugs and Biologicals, published by the Merck Company, Rahway, NJ. Gives a concise summary of the physical and biological properties as well as uses, toxicity and hazards of more than 10,000 compounds, with some literature references. All compounds including synonyms and trade names are alphabetically order. This handbook has a formula index and subject index.

③ Beilstein Handbook of Organic Chemistry. Beilstein handbook is a chemical information system for organic compounds consisting of structure, property, reaction and bibliographic databases covering over 7 million substances and 5 million reactions. It was originally published in German, but it is now published in English. Records in the database include structures, data on chemical and physical properties such as melting point, boiling point and spectral identification, as well as reference to the chemical literature published from 1979 to the present. The search can also be done online.

(2) Journals

Main research journals include:

①《Journal of the Chemical Society (Perkin Transactions)》Published by the Royal Society of Chemistry (UK). Mainly devoted to organic synthesis

②《Journal of the American Chemical Society》. It is the mot important and prestigious journal in chemistry.

③《Journal of Organic Chemistry》. Published by American Chemical Society. Includes all areas of organic chemistry.

④《Journal of Organic Chemistry》. Also published by American Chemical Society. Comprehensive reviews on chemistry

⑤《Tetrahedron》and《Tetrahedron letter》. The first is full articles about organic chemistry, and the second is for timely publication of short communication, most of them related to organic synthesis.

⑥《Journal of Chemical Education》. Published by the Division of Chemical of the

American Chemical Society. Mainly devoted to chemical education.

(3) Chemical Abstracts

The Chemical Abstracts, published by the American Chemical Society, is the most complete chemical information resource. It began in 1907 and annually publishes abstracts from about 9, 000 journals as well as books, conference proceedings, dissertations and patents. Chemical Abstracts has several indexes including chemical substance, general subject, author, formula and patent number, all of which are accumulated at five-year intervals. It provides concise summaries of original research articles in journals and listings of reviews and books. Also, it is possible to search Chemical Abstracts online.

(4) Online Resources

Searching for information online is more and more convenient, with the fast growth of the internet. There are many ways to get chemical resources from the internet

① Online Libraries

http: //www. nlc. gov. cn/

http: //www. lib. tsinghua. edu. cn/

http: //www. lib. pku. edu. cn/html

② Online Periodical Resources

http: //www. chinajournalnet/

http: //www. cqvip. com/

③ Online Patent Resources

http: //www. patents. lbm. com/

http: //www. patent. com. cn/

http: //www. cnpatent. com/

④ Database Resources

http: //www. colby, edu/chemistry/cmp/cmp, html

http: //www. chemfinder. camsoft. com/ (basic property of compounds)

http: //www. sioc. as. cn/lccdeb/sjk. htm (IR spectrum database)

⑤ Other Website

http: //www. acs. org/

http: //www. ccs. ac. cn/

http: //www. chemsoc, org/ (ChemSoc)

http: //www. csir. org/ (Chemistry Software and Information Resources)

http: //jchemed. chem. wise. edu.

http: //www. chemistry. rsc. org. rsc/ (J. Chem. Soc)

http: //www. chemcenter. org/ (J. Amer. Chem. Soc)

Reference

[1] L. M. Harwood, et al. Experimental Organic Chemistry, 2nd. ed. Oxford: Blackwell Science Ltd. (add: Osney Mead, Oxford OX2 0EL), 1999.

[2] Xi Guan Gen, et al. Experimental Organic Chemistry. Shanghai: East China University of Science and

Technology Press，1999.

[3] Zeng Zhao Qiong，et al. Experimental Organic Chemistry，2nd. Peking：Higher Education Press，1987.

[4] Royston M Roberts，etal. Modern Experimental Organic Chemistry. New York：CBS College Publishing，1985.

[5] Daniel R. Palleros. Experimental Organic Chemistry. New York：Jony Wiley & Sons，Inc，2000.

[6] Charles F. Wilcox，Jr. Experimental Organic Chemistry. MacMillan Publishing Company，a division of MacMillan，Inc，1998.

[7] John C. Gilbert，Stephen F. Martin. Experimental Organic Chemistry，2nd，ed. New York：Saunders College Publishing，1998.

Chapter 2 Basic Experimental Techniques

2.1 Heating and Cooling

Heating and cooling is one of the basic techniques organic chemistry laboratory.

2.1.1 Heating

There are several common heat sources such as gas, alcohol and electric power in the organic chemistry laboratory. Heating methods can be divided into direct and indirect ones. To avoid possible problems from straightforward heating, use the following alternate heating methods when necessary.

Air-bath: With the principle of indirect heating by hot air, liquids can be heated above 80℃. The most common air bath in the experiment is heating with a wire gauze and a Bunsen burner or with an electric heating mantle, which is designed only for heating round-bottom flask.

Water bath: Usually a heated water bath is a convenient way of heating a liquid below 80℃. During this operation, the vessel must be immersed into a water bath so that the surface of water bath is kept higher than inside surface of the solution.

Oil bath: The usable range of an oil bath is 100～250℃. The highest temperature that oil bath can reach is depended upon what material is used. The plant oil used in the lab can reach 200～220℃, and liquid wax can reach 220℃. Pay attention to the safety when heating an oil-bath to prevent accidental fires or splattering from water introduced into the hot oil.

Sand bath: When a heating temperature is required to get higher than those listed above, one can often use a sand bath. Its highest operating temperature is 350℃.

2.1.2 Cooling

Cooling is also needed when the organic experiment has to be run at low temperature use the following cooling methods, based on different requirements.

For ordinary cooling, the container with reactant must be put into cold water. Choose ice or an ice-water mixture for cooling below room temperature. For cooling below 0℃, use broken ice and some inorganic salts mixed with certain ratios as a cooling agent. Solid CO_2 mixed respectively with acetone, chloroform or other solvents can reach −78℃. Liquid nitrogen can be as cold as −196℃.

When the temperature is below −38℃, don't use a mercury thermometer but an ethanol-based low temperature thermometer.

2.2 Drying and Drying Agent

Drying is a very useful technique in organic experiments. The common drying method can be primarily divided into physical and chemical ones. The physical drying method includes absorption, dehydration with molecular sieves, etc. The chemical drying method is to use a drying agents to remove water. The latter based on the chemical reaction with water of the drying agents to form hydrated compounds.

2.2.1 Drying a Liquid Organic Compound

2.2.1.1 Selection of Drying Agent

Usually a small portion of the desired drying agent is added directly to an organic solution. Four points must be followed when selecting a drying agent: (a) there should be no reaction with the organic compound to be dried; (b) it should be insoluble in organic compound to be dried; (c) it should have a large capacity of water uptake; (d) it should absorb water quickly and be low cost. The properties of common drying agents are shown in Table 2.1.

Table 2.1 The properties of Some Drying Agents

Drying agents	Capacity	Intensity	Speed
$CaCl_2$	0.97($CaCl_2 \cdot 6H_2O$)	Medium	Very fast, but stand longer
$MgSO_4$	1.05($MgSO_4 \cdot 7H_2O$)	Weak	Very fast
Na_2SO_4	1.25	Weak	Slow
Ca_2SO_4	0.06	Strong	Fast
K_2CO_3	0.2	Quick weak	Slow
CaO	—	Strong	Very fast
P_2O_5	—	Strong	Fast
Molecular sieves	About 0.2	Strong	Fast

2.2.1.2 Dosage of Drying Agent

The proper dosage of a drying agent, usually more than its theoretical value, can be figured out, based on its capacity to take up water and water solubility in the target organic solvent. In view of molecular structure, slightly move drying agent can be used for those compounds containing the water-philic group. But the drying agent must be used appropriately. The drying of compound is incomplete if there is insufficient drying agent, but loss of the organic compound results from too much drying agent due to surface adsorption. So the general dosage 0.5～1g dry agent per 10ml liquid (see Table 2.2).

Table 2.2 Some Common Drying Agents for Organic Compounds

Organic compounds	Drying agent	Organic compounds	Drying agent
Hydrocarbon	$CaCl_2$, Na, P_2O_5	Ketone	K_2CO_3, $CaCl_2$, $MgSO_4$, Na_2SO_4
Halohydrocarbon	$CaCl_2$, $MgSO_4$, Na_2SO_4, P_2O_5	Acid, Phenol	$MgSO_4$, Na_2SO_4
Alcohol	K_2CO_3, $MgSO_4$, CaO, Na_2SO_4	Ester	K_2CO_3, $MgSO_4$, Na_2SO_4, K_2CO_3
Ether	$CaCl_2$, Na, P_2O_5	Amine	KOH, NaOH, K_2CO_3, CaO
Aldehyde	$MgSO_4$, Na_2SO_4	Nitrocompound	$CaCl_2$, $MgSO_4$, Na_2SO_4

2.2.1.3 Methods of Drying

Before drying, remove water from organic compounds as much as possible, using a separatory funnel if necessary. Then add a small amount of drying agent directly to the organic solution. Swirl the solution and let it stand while you observe the drying agent. If it is all clumped together, add more and observe how a solution with drying agent looks when it is clumped and when it is free-flowing. Continue swirling and observing the solution for several minutes, adding more drying agent only until a fresh addition no longer forms clumps. In most cases, the particle size of drying agent should be handpicked suitably. The larger the particle size, the slower the speed of drying; the smaller the particle size, the more organic substances can be absorbed onto the surface.

2.2.2 Drying Solid Organic Substances

Air dry: Spread solid samples out on a watch glass or filter paper and allow them to air dry.

Oven dry: You can dry solid samples more quickly by putting them on a hot-water-bath or in a large beaker in an oven at a temperature below their melting point in order to avoid their decomposition.

Other drying methods: Infrared desiccator, vacuum desiccator, etc.

2.3 Filtration

Filtration is solid-liquid separation method by porous media (such as filter paper, filter cloth) to block the solid particles of the suspension solid-liquid. Commonly, the filtering methods are pressure filtration, vacuum filtration, filtering hot, then solid materials to remain in the filter paper, the liquid filtration goes through the filter into flows the receiving container, the solution obtained by filter is called filtrate.

Temperature, viscosity and filtration pressure of solid-liquid mixture and the state of material will affect the speed of filter. Hot solution is easily filtered than cold one; the greater viscosity of the solution, the filter will be more slowly; decompression filter is faster than atmospheric pressure; it must be destroyed by heating when it is rubbery solid, to avoid it through filter paper. In a word, it is necessary to consider various factors when choosing different filtering methods.

2.3.1 **Atmospheric Pressure Filtration** (Common Filtration)

The method of atmospheric pressure filtration is the filter method by using common filter funnel under atmospheric pressure. The driving force of filter is gravity. So it is the most simple and general, but its peed is slowly.

According to the nature of precipitation to choose the type of filter paper: we will choose slow filter paper that fine crystal precipitation, we will choose fast filter paper in colloidal precipitation, we will choose medium-speed filter paper in coarse crystal precipitation. We will select the size of filter paper in accordance with the size of filter funnel.

The operation of atmospheric pressure filtration is shown in Figure 2.1.

This method is commonly used in the removal of insoluble substances. Sometimes it can be used to separate crystals from the larger particles, absorbent poor crystal products. The method also is used in analytical chemistry.

2. 3. 2 Decompression Filtration (Vacuum Filtration)

Decompression filtration is rapid to take away air from filter bottle by using the suction pump, so through creating the pressure difference of Buchner funnel surface and filter bottle, the filtration is speeded up. Safety bottle is assembled between the pump and suction filter bottle, to prevent the phenomenon of reverse suction because closing pump. This is apparatus of decompression filtration (see Figure 2. 2).

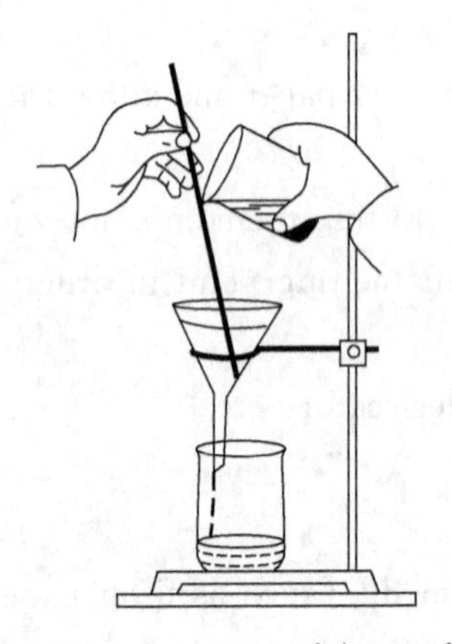

Figure 2. 1 Apparatus of Atmospheric Pressure Filtration

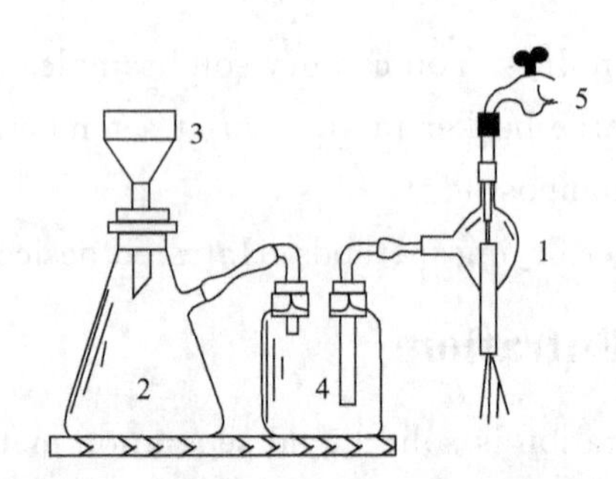

Figure 2. 2 Apparatus of Decompression Filtration

1—Pump; 2—filter bottle; 3—Buchner funnel; 4—safety bottle; 5—faucet

You must pay attention that filter paper should be slightly smaller than the diameter of Buchner funnel, but able to cover all porcelain hole when the method is used. Opening pump after filter paper wetted with distilled water, then transferring samples. You should open the piston of safety bottle and then close pump at the end of suction filtration.

This method of decompression filtration is unsuitable for filtration of the precipitate colloidal and smaller particles.

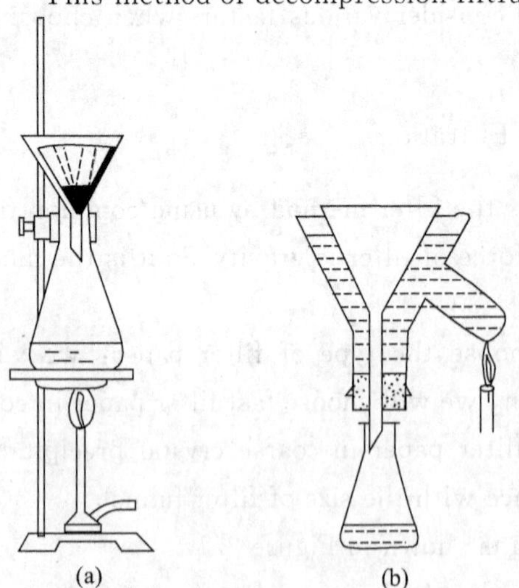

Figure 2. 3 Apparatus of Hot Filtration

2. 3. 3 Hot filtration

If it is obvious that the solubility of material in the solvent changes with temperature, hot filtration is necessary, to prevent the precipitation of crystals in the filter paper. See Figure 2. 3.

Filter paper was folded into a folded filter paper (see Figure 2. 4), in order to increase the filter area to speed up the filtration rate. The transferring solution is rapid when filtration, then the funnel was covered with the concave downward watch glass to prevent solvent volatile.

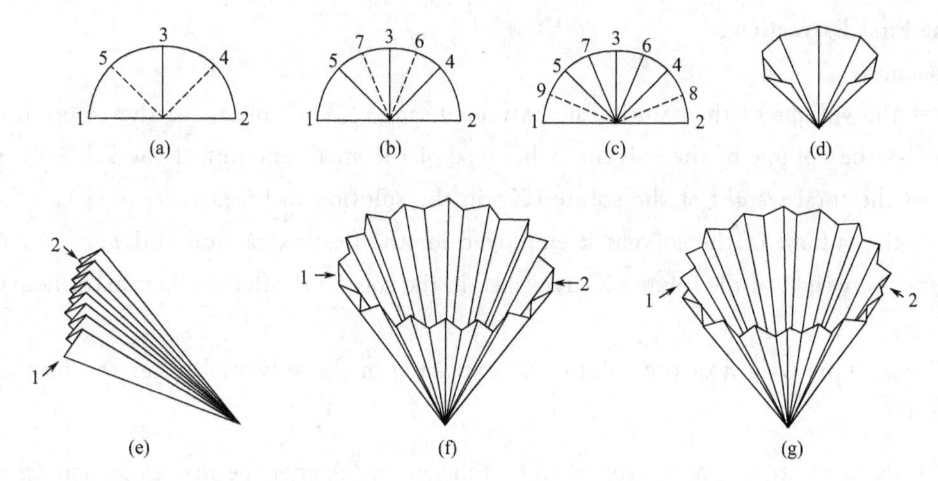

Figure 2.4 Folding Filter Paper to Produce Fluted Filter Paper

2.4 Solvent Extraction and Solution Washing

2.4.1 Principle

Solvent extraction is technique frequently used in the organic chemistry laboratory to separate or isolate a desired species from a mixture of compounds or from impurities. Substances for extraction usually are solid or liquid. The apparatus for extraction from liquid is called separatory funnel, and its use is one of the essential operations for organic chemistry experiments. Solution washing is the reverse process of solvent extraction. Usually it is performed in the case of impurities in a solution that can be extracted by another immiscible solvent, which has a small solubility to the desired substance. Separatory funnels are also the essential apparatus for solution washing, and the principle is similar to that of solvent extraction. Here we will focus our discussion on solvent extraction.

The extraction principle can be illustrated by the following calculation. It is assumed that the solution is comprised of an organic compound X dissolved in a solvent A. If we will extract X from A, we can choose a solvent B, which has a higher solubility to X than A, and it is immiscible and dose not react with B. Place the solution into a separatory funnel, then add B, shake or swirl the funnel and stand the funnel upright until a sharp demarcation line appears between the two layers because of the immiscibility of B with A. The concentration ratio of X in solvent B and A is a constant at a given temperature, and the ratio is also called the distribution coefficient, expressed as K. This relationship is known as distribution principle, which can be expressed by the following equation:

$$\frac{\text{Concentration of X in the solvent A}}{\text{Concentration of X in the solvent B}}=K$$

According to the distribution principle, in order to economize solvent and improve extraction efficiency, multiple extractions with small volumes of extracts are more efficient than a single extraction with a large volume.

The First Extraction

Assume:

V = the volume of the solution be extracted (mL) (The volume of the solution can be regarded as the volume of the solvent A because of the small amount of the solute X);

m_0 = the total weight of the solute (X) in the solution to be extraction (g);

S = the volume of the solvent B employed for the first extraction (mL);

m_1 = the weight of the solute (X) remained in the solvent A after the first extraction (g).

Then:

$m_0 - m_1$ = the weight of the solute (X) extracted in the solvent B after the first extraction (g);

$\frac{m_1}{V}$ = the concentration of the solute (X) in the solvent A after the first extraction (g/mL);

$\frac{m_0 - m_1}{S}$ = the concentration of the solute (X) in the solvent B after the first extraction (g/mL);

From $\frac{\frac{m_1}{V}}{\frac{m_0 - m_1}{S}} = K$, we can obtain the result: $m_1 = m_0 \frac{KV}{KV+S}$

The Second Extraction

Assume:

V = the volume of the solution be extracted (mL);

m_2 = the weight of the solute (X) remained in the solvent A after the second extraction (g);

S = the volume of the solvent B used in the second extraction (mL).

Then:

$m_1 - m_2$ = the weight of the solute (X) extracted in the solvent B after the second extraction (g);

$\frac{m_2}{V}$ = the concentration of the solute (X) in the solvent A after the second extraction (g/mL);

$\frac{m_1 - m_2}{S}$ = the concentration of the solute (X) in the solvent B after the second extraction (g/mL).

So:

$\frac{\frac{m_2}{V}}{\frac{m_1 - m_2}{S}} = K$, we will get the result $m_2 = m_1 \frac{KV}{KV+S}$.

Replace m_1 with $m_0 \frac{KV}{KV+S}$ (obtained from the first extraction), the above relationship can be rewritten as:

$$m_2=m_0\left(\frac{KV}{KV+S}\right)^2$$

For more times of extractions, they can be done in the similar way. If we assume the employed volume of the solvent B each time, and m_n equals the weight of the solute (X) remained in the solvent A, after multiple (n) extractions (g), we can obtain m_n, from the following equation:

$$m_n=m_0\left(\frac{KV}{KV+S}\right)^n$$

2.4.2 Procedure

2.4.2.1 The Extraction of Liquid-liquid

The extracting from the liquid (or washing) is usually carried out through separatory funnel. There are three kinds of separatory funnel-spherical, subulate and pear forms. In organic experiments, a separatory funnel is mainly applied to:

a. Separate two kinds of liquid which are immiscible and do not react with each other;

b. Extract a certain component from a solution;

c. Wash a liquid product or the solution of a solid product with water, aqueous alkaline or acidic solution;

d. Add dropwise a certain reagent (used as a dripping funnel).

The suitable volume of the separatory funnel is more than double of the volume of extracted solution. You should check whether leakage of the plug and piston with water before experiment.

The separatory funnel should be placed inside the ring of the iron support stand when extraction (or washing), closing the pistons, add them into the extracting solution and reagents. Insert the stopper, then pick up the funnel in both hands and invert it with the one hand holding the stopcock and the first two fingers of another hand holding the stopper in place. Shake or swirl the funnel gentle for a few seconds, vent the funnel slowly by opening the stopcock to release any pressure buildup. Close the stopcock and shake the funnel more vigorously, with occasional venting, for 2～3 minutes, so that the two immiscible solvents can contact with each other to increase efficiency. (see Figure 2.5)

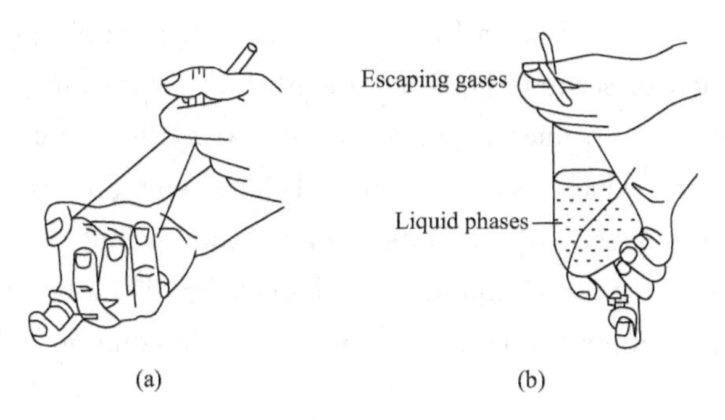

Figure 2.5 Venting Separatory Funnel

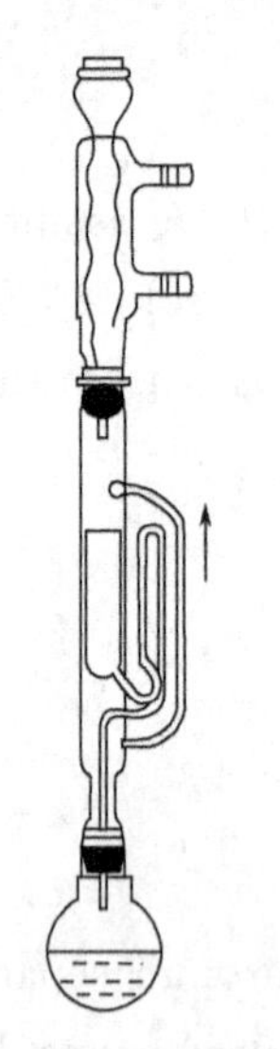
Figure 2.6 Soxhlet Extractor

Replace the funnel on the ring, remove the stopper, and allow the funnel to stand until there is a sharp demarcation line between the two layers. Drain the bottom layer into an flask by opening the stopcock fully; turn it to slow the drainage rate as the interface approaches the bottom of the funnel. When the interface just reaches the outlet, quickly close the stopcock to separate the layers cleanly. Repeated the above operation.

2.4.2.2 The Extraction of Liquid-solid

For a solid mixture, liquid-solid extraction can be performed with a Soxhlet extractor (see Figure 2.6). The solid is placed in a porous thimble. The extraction-solvent vapor, generated by refluxing the extraction solvent contained in the distilling pot, passes up through the vertical side tube into the condenser. The liquid condensate then drips onto the solid, which is extracted, The extraction solution passes through the pores of the thimble, eventually filling the center section of the Soxhlet. The siphon tube also fill with extraction solution and when the liquid level reaches the top of the tube, the siphoning action commences and the extract is remained to the distillation spot. The cycle is automatically repeated numerous times. In this manner the desires species is concentrated in the distillation pot. Equilibrium is not generally established in the system, and usually the extraction effect is very high. After the liquid-solid extraction, the solution can be treated as usual to obtain the desired product.

2.5 The Purification and Separation of Liquid Organic Substances

2.5.1 Simple Distillation

2.5.1.1 Principle

Distillation has been used for centuries in the purification and separation of liquid organic substances. When the vapor pressure of the liquid being heated equals the atmospheric pressure on the surface of the liquid, it boils. The temperature, at that moment, is called boiling point. Simple distillation is a process that consists of boiling a liquid and condensing the vapor to the liquid state.

Every pure organic compound has a fixed boiling point at a certain pressure. Whenever possible, use simple distillation to separate liquid mixtures which have boiling points more than about 30℃ apart. But sometimes an organic compound together with other components can form a binary or ternary azeotrope mixture which also has a definite boiling point. So it is difficult to say whether a liquid is a pure organic compound, based on its definite boiling point.

The boiling range of a pure compound is about 0.5～1℃, but that of mixture is larger. Usually you can determine the boiling point of a liquid compound by distillation.

2.5.1.2 Procedure

① Assemble a simple distillation apparatus shown in Figure 2.7 according to the "bottom-to-top" and "left-to-right" principle. Make sure all taper joints are connected securely. Select a round-bottom flask of such a size that it should be no more than one-half full. Insert a thermometer into the distillation head through a thermometer adapter with the top of the mercury bulb lined up with the bottom of the side-arm of head so that the bulb will be bathed in distillate vapors just before they enter the condenser.

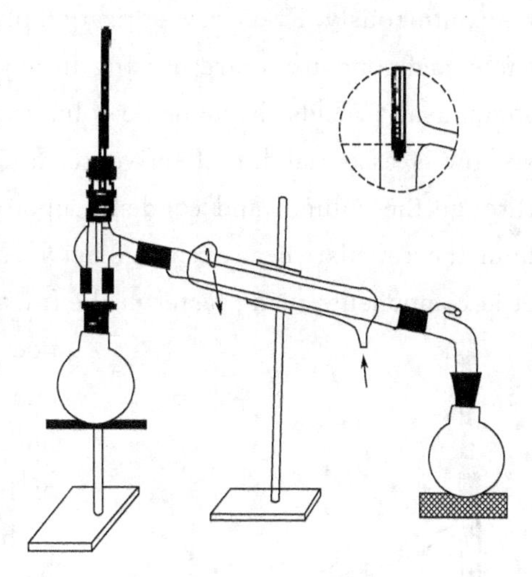

Figure 2.7 Apparatus of Simple Distillation

② Put the liquid into the flask through a separatory funnel, add a couple of boiling chips[1], and introduce the cooling water into a condenser[2].

③ First heat slowly with a heating mantle, and then gradually increase the power, make the liquid boil, and start distilling[3].

④ Adjust the heating source, maintain distillation at a rate of 1~2 drops/s, and record the temperature of first distillate (maybe forerun, if any).

⑤ Keep heating speed constant, continue to distill, and collect all distillates over the range of the desired temperature. Discontinue distilling any liquid[4] when no distillates come out and the temperature decreases suddenly.

⑥ Remove the heating source, stop the cooling water, and disassemble the apparatus following the "right-to-left" and "top-to-bottom" principle.

Notes

[1] Adding boiling chips can produce smooth bubbling and prevent boil-over or bumping of the liquid. Avoid adding them to a heated liquid. If distilling is started after heating is stopped.

[2] Introduce tap water from a hose into the lower condenser nipple and have it exit of the upper nipple.

[3] The temperature at this moment is the boiling point of the distillate.

[4] Don't distill to dryness to prevent any experimental accidents.

2.5.2 Fractional Distillation

2.5.2.1 Principle

Fractional distillation, more effective than simple distillation, is also a commonly used technique of the separation and purification of liquid organic compounds with a smaller boiling point difference.

Fractional distillation is a distillation operation with a fractionating column which is equivalent to several successive simple distillations. As the vapor from the distillation flask rises up through the fractionating column, it condenses on the column packing and revaporizes continuously. Every revaporization of the condensate is just a separate distillation. In fact, vapor and condensate are passing in opposite directions through the column; more volatile component ascends through the column, while less volatile components descend. The counter-flow is essential for effective separation in a fractionating column. The vapors pass up through the column and condense upon heat exchange with the cooler surface of the column. There exists heat exchange between vapor and condensate. This results in the more volatile components rising faster in the fractionating column. If repeated, that means the process goes through many vaporization- condensation cycles and has the effect of many distillations. So there is a high ratio of the more volatile components near the top of fractionating column, and on the other hand, the high boiling-point components are moving downward to the distillation flask. When the column efficiency is high enough, the vapor out of the top of the fractionating column is close to the pure low-boiling component; meanwhile, the high boiling-point components are in the flask. Thus, components with small differences in boiling point can be separated.

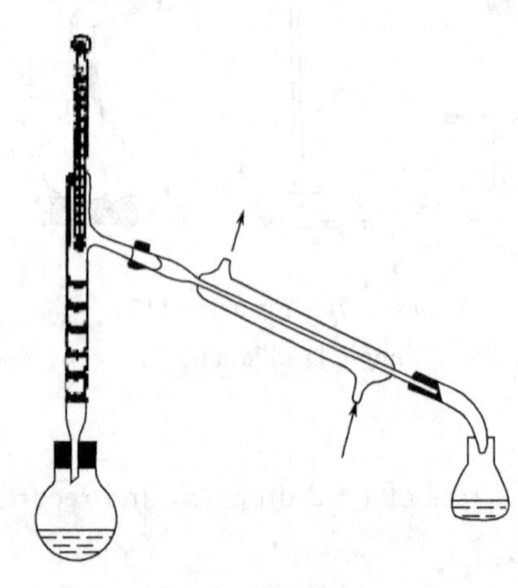

Figure 2.8 Apparatus of Fractional Distillation

2.5.2.2 Procedure

① Select a fractionating column, assemble an apparatus of fractional distillation shown in Figure 2.8. Make sure all joints are connected well.

② Put the liquid to be distilled into a flask, add a couple of boiling chips, and introduce tap water into the condenser.

③ Control heating speed, maintain a takeoff of no more than 1～2 drops/s[1].

④ Collect distillates at different temperature ranges[2], record the temperature of the different distillates respectively.

Notes

[1] If the distillation rate is too quick, the purity of products will decrease. If the distillation rate is too slow, the distillation temperature fluctuates up and down. To reduce the heat loss in a fractionating column, usually use asbestos rope to wrap it.

[2] Don't distill to dryness.

2.5.3 Steam Distillation

2.5.3.1 Principle

Steam distillation, is one of the most commonly used methods of separating and purifying liquids. In this method steam is passed into the insoluble or immiscible and volatile liquid in the distillation flask, and then the water and organic compound co-distill, condense in the condenser, and are collected in the normal way.

According to the law of partial pressure above the mixture is the sum of the partial pressures of the pure individual components. Therefore, when the sum of the individual vapor pressures is equal to the atmospheric pressure, the mixture will start boiling at a lower temperature (boiling point) than either of the individual components. That means that the organic compound will distill out below its normal boiling point.

In the distillate, the distillate, the weight ratio of organic compound (m_A) and water (m_{H_2O}) can be expressed as in the following equation.

$$\frac{m_A}{m_{H_2O}}=\frac{p_A M_A}{p_{H_2O} M_{H_2O}}$$

Steam distillation is applicable to the following situation: ①a high boiling-point organic compound which decomposes at its boiling point; ②a mixture containing inorganic salts or other organic solids, not suitable for distillation, filtration and extraction; ③a mixture containing large amounts of resin-like, nearly intractable substances or non-volatile impurities which is hard to isolate or purify in any other way.

In order for steam distillation to be feasible, the substances to be purified should have these characteristics: ①insoluble or immiscible in water; ②no reaction with water when boiling; ③having a certain vapor pressure at 100℃.

2.5.3.2 Procedure

① Assemble an apparatus shown in Figure 2.9, make sure to fasten securely and smoothly steam lines at all keep the lines from the steam generator to three-neck flask as short as possible. they are connected by a T-shaped tube. Underside of the T-shaped tube is connected by a short rubber tube with the screw clamp.

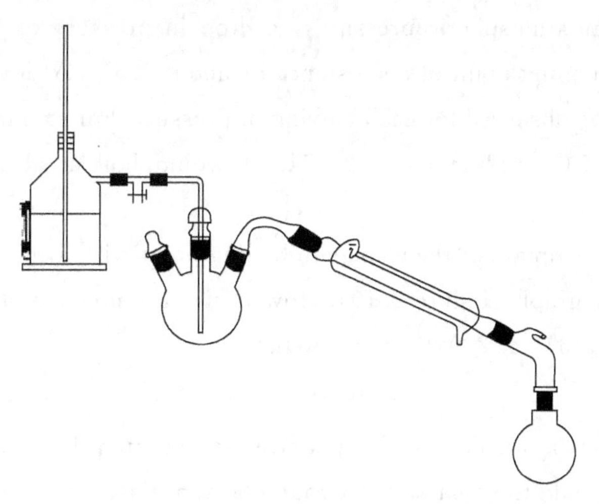

Figure 2.9 Apparatus of Steam Distillation

② Pour the distillate into the round-bottom flask. Be certain that the flask is filled to no more than one-third of its capacity. Make sure that all connection are tight.

③ Open the screw clamp on the T-shaped tube connecting the three-neck flask and the vapor generator before distillation. Introduce tap water to the condenser. Heat the steam generator. When steam goes out of vapor generator and beings to enter the flask, close the clamp up[1].

④ Adjust a heating source and make sure that the rate of distillate entering the receiver is at 1～2 drops/s[2].

⑤ When oily droplets are no longer observable or the distillate turns transparent or clear, stop heating immediately.

⑥ Open the clamp, remove the heat source[3], disassemble the apparatus, in turn.

Notes

[1] At this moment the solution in the flask bubbles furiously, the mixture of organic compound and water distills off soon.

[2] In order to avoid filling with condensed water in the three-neck flask during distillation, heat the flask with a small flame to make the on-going distillation proceed more quickly. Watch the water level height in the safety column at all times. If abnormal, open the clamp immediately, remove heat source, get rid of any troubles, and then go ahead.

[3] Avoid letting the liquid from the flask back up into the steam line.

2. 5. 4 Vacuum Distillation

2. 5. 4. 1 Principle

When the pressure over a liquid is reduced, the liquid boils at a lowered temperature. As its name implies vacuum distillation under reduced pressure. As a rule, most liquids that boil around 200℃ or above at normal atmospheric pressure can be purified by vacuum distillation, since above that temperature, chemical transformations are more likely.

At pressures near atmospheric pressures, a drop in pressure of 10mmHg (1. 33kPa) generally lowers the boiling point of a substance by about 0. 5℃. At lower pressures, a 10℃ drop in boiling point is observed for each halving of pressure. For example, a compound with a boiling point of 100℃ at 20mmHg (2. 67kPa) would boil at about 90℃ at 10mmHg (1. 33kPa).

A more accurate estimate of the change in boiling point with a change of pressure can be made by using a nomograph (Figure 2. 10). How to use the nomograph: assume a reported boiling point will be 113℃ at 2. 0kPa (15mmHg).

2. 5. 4. 2 Apparatus

Figure 2. 11 shows a standard scale apparatus for vacuum distillation. The main equipment and supplies include the heat source, round-bottom flask, claisen head, capillary bubbler, thermometer, condenser, vacuum adapter, receiving flask, trap, manometer and vacuum sources (a water aspirator or a vacuum pump).

The principal difference between a vacuum and an atmosphere distillation set-up is that in vacuum distillation, the system must be air tight and able to hold a vacuum. The glassware used to assemble them must be able to withstand the vacuum; for example, large, thin-walled flasks cannot be used. Use vacuum grease to seal all glass-to-glass connections. Stopcock grease flows too easily under heat to hold a vacuum. Use only heavy-walled vacuum tub-

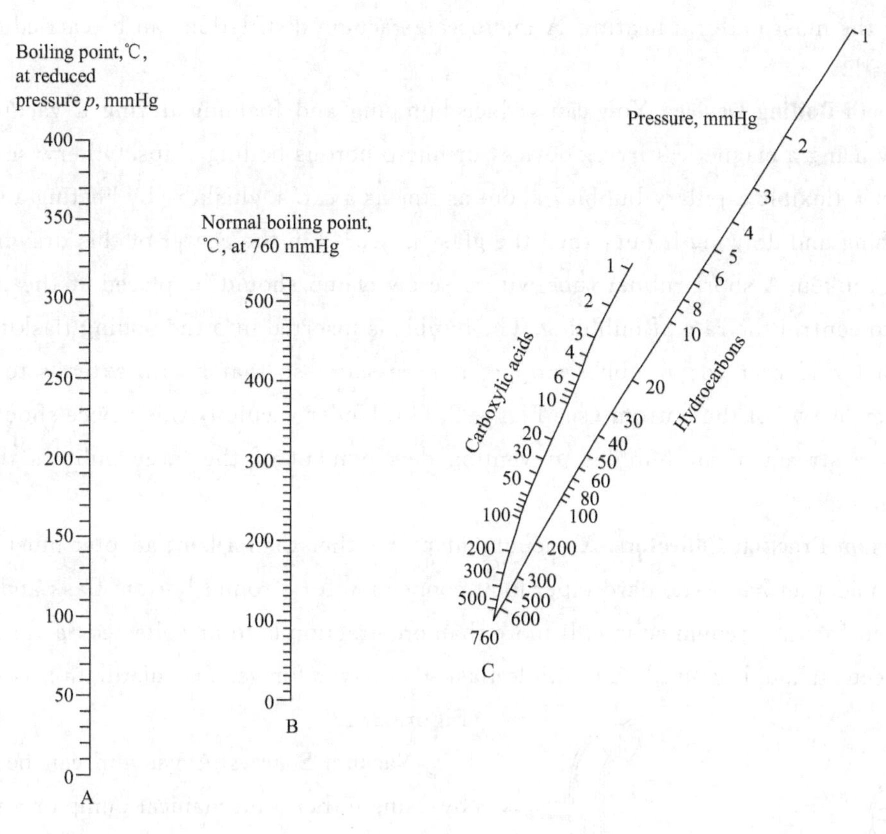

Figure 2.10 A Nomograph for Estimating Boiling Points at Different Pressures

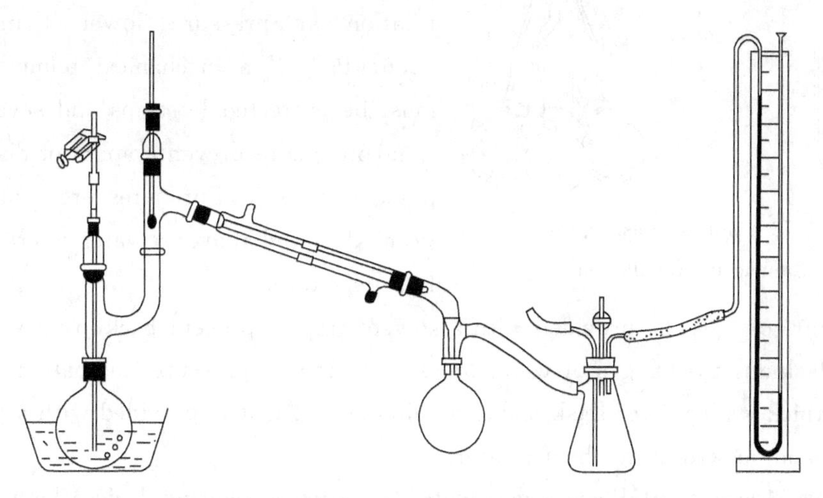
Figure 2.11 Apparatus for Vacuum Distillation

ing to prevent collapsing under vacuum. Connections between rubber and glass, as well as ground-joint connections, must be secure and airtight.

Under vacuum, most solvents boil well below room temperature; therefore, any solvent in the material to be distilled must be removed prior to distillation.

Heat Sources. The heat source should be capable of providing constant uniform heating to prevent bumping and superheating and to maintain a constant distillation rate. An oil bath provides the most uniform heating. A microscale vacuum distillation can be carried out using a sand bath.

Smooth-Boiling Devices. You can reduce bumping and foaming during a vacuum distillation by using a magnetic stirring device, or micro-porous boiling chips. Otherwise, you can construct a flexible capillary bubble, about as fine as a cat's whisker, by heating a section of glass tubing and drawing it out, then the glass is scored in the center of this drawn-out section and broken. A short rubber tube with a screw clamp should be placed at the top of the bubble to control the rate of bubbling. The bubble is inserted into the boiling flask through a thermometer adapter (or a rubber stopper if necessary) so that its tip extends to within a millimeter or two of the bottom (see Figure 2. 11). Under vacuum, this device should deliver a very fine stream of air bubbles preventing development of the large bubbles that cause bumping.

Vacuum Fraction Collectors. A vacuum adapter rather than a bent adapter must be used, since this adapter has a standard-taper joint connection for a round-bottom flask and a tubing connection for the vacuum source. If more than one fraction is to be collected, a vacuum fraction collector should be used. A multiple-flasked receiver for vacuum distillation is shown in Figure 2. 12.

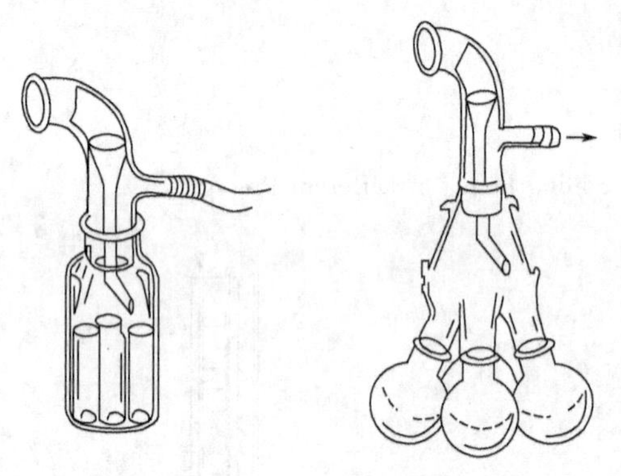

Figure 2. 12 Multiple-flasked Receiver for Vacuum Distillation

Vacuum Sources. A vacuum can be obtained by using either a mechanical pump or a water aspirator. A mechanical pump must be used for distillation at pressure lower than 20mmHg (2. 67kPa). If a mechanical pump is used, it must be protected by traps and several absorption columns to prevent vapors of distillate from passing into it. For pressures of 20mmHg (2. 67kPa) and above, a water aspirator is usually satisfactory.

An aspirator must be provided with a solvent trap to prevent backup of water into the receiving flask due to changes of water pressure to reduce pressure fluctuations throughout system. A thick-walled filter flask makes a suitable trap if it is provided with a pressure release valve and hooked up to the aspirator.

Pressure Measurement. Place a manometer (pressure-measuring device) between the vacuum source and the vacuum adapter. Two kinds of closed-end manometers are suitable for measuring pressure. In either of these manometers, the unit difference in the levels of mercury in mm, is equal to the pressure in the system. The mercury in a manometer can present a hazard if the vacuum is broken suddenly—air rushing in can push the mercury column forcefully to the closed end of the tube, breaking it and releasing toxic mercury into the laboratory. For many reasons, the vacuum must always be released slowly. It is advisable to open the manometer to the system only when a pressure reading is being made.

2.5.4.3 Steps in Vacuum Distillation

① Assemble the apparatus illustrated in Figure 2.11, applying vacuum grease at all joints. With an empty distillation flask, test the seals by applying vacuum after closing the screw clamp at the top of the capillary tube and opening the stopcock of the safety flask. You should be able to obtain a pressure of approximately 20mmHg (2.67kPa) using an aspirator or a much lower pressure with a mechanical pump. If you cannot obtain this low pressure, you should use the pinch clamp to isolate various parts of the system to find the leak (s).

After checking the empty system, break the vacuum by opening the stopcock of the safety flask slowly while watching the mercury in the manometer rise.

② Using a funnel with a stem, pour the liquid to be distilled into a distillation flask until the flask is no more than half-full. Add a few micro-porous boiling chips if you are not using a capillary tube. After closing the stopcock of the safety flask, apply a full vacuum with the aspirator or the mechanical pump. Adjust the screw clamp until a fine stream of bubbles appears from the bottom of capillary tube. You should be able to attain nearly the same low pressure as when the distillation flask was empty.

③ Adjust the pressure and let the system stand for a few minutes until the pressure is no longer changing. Watching the manometer, adjust the clamp if you want to conduct the distillation at a specific pressure, other than maximum attainable vacuum. Turn on the condenser cooling water, heat the liquid, with an oil bath and distill it. Record the temperature and pressure, then check them occasionally. If pressure changes, adjust the heat input accordingly. The distillate should be collected at a rate of about 1～2drops/s.

④ At the end of the distillation, removed the heat source, and slowly open the screw clamp and the stopcock of the safety flask to release the vacuum. Then turn off the aspirator or the mechanical pump.

⑤ If a forerun must be removed, or if you are collecting fractions, you should break the vacuum and change the receivers. A more convenient way to collect fractions is to use a multriple-flasked receiver (Figure 2.12). With a multriple-flasked receiver you can collect several fractions before breaking vacuum. You must cool the flask when sudden violent boiling or forcing might occur after you reestablish the vacuum.

⑥ Remove the receiving flask and clean all glassware as soon as possible after disassembly to keep the ground glass joints from sticking.

2.5.5 Rotatory Evaporator

Rotary Evaporator operation is the method of separating organic solvents. Rotary evaporator operation is completed in the rotary evaporator. The structure of the rotary evaporator is shown in Figure 2.13. It consists of rotating evaporator driven by a motor, condenser, collector. It can be used under normal pressure or decompression. As a result of constantly rotating evaporator to avoid bumping. Meanwhile liquid is attached to the wall to form a layer of liquid film and increase the evaporation area, so evaporating faster. You should first stop the heating and vacuum pump and finally cut off the power when finishing

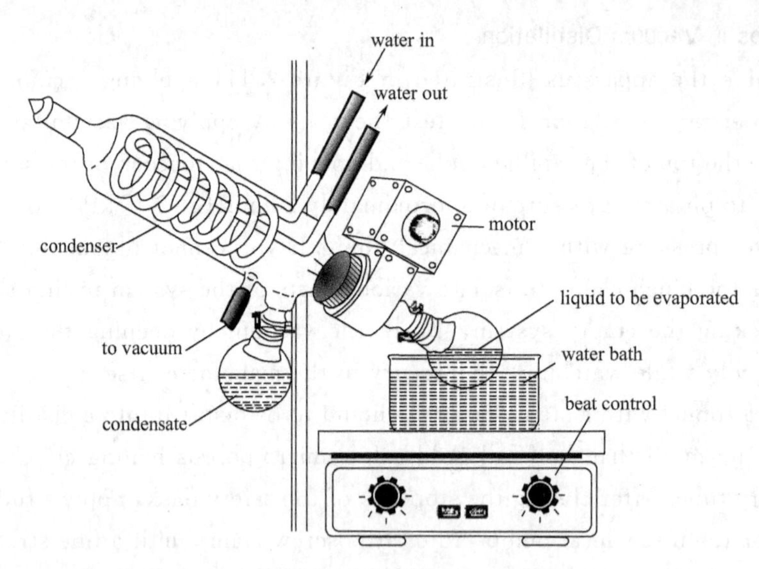

Figure 2. 13 Apparatus of Rotatory Evaporator

evaporation.

2.6 The Purification and Separation of Solid Organic Substances

2.6.1 Recrystallization

2.6.1.1 Principle

Recrystallization is the most frequently used operation for purifying organic solids. This technique is based on the fact that the solubility of an organic compound in a given solvent will often increase greatly as the solvent is heated to its boiling point. When it is first heated in such a solvent until it dissolves, then cooled to room temperature or below, an impure organic solid will usually recrystallize from solution in a much purer form than its original form. Most of the impurities will either not dissolve in the hot solution (from which they can be filtered) or remain in dissolved form in the cooled solution (from which the pure crystals are filtered).

2.6.1.2 Procedure

(1) The General Procedures for Recrystallization

① Select a suitable solvent.

② Dissolve the crude product in the hot solvent and make it saturate.

③ Filtrate off undissolved impurities. If the solution shade is deep, decolorize it and then do filtration.

④ Cool down the solution or evaporate the solvent to make the crystals precipitate. The impurities are left in the solution or are crystallized out and the desired product is left in the solution.

⑤ Collect the crystals by filtration, rinse the crystals and remove the last traces of solvent.

⑥ Dry the crystals.

(2) The Choice of Solvent System

In the operation of recrystallization, the choice of solvent is critical to achieve high purity and recovery. As a suitable solvent, the following requirements should be met:

① Does not react with the substance to be purified.

② Possesses high solubility if the solute at elevated temperature and low solubility at low temperature.

③ If the impurities are insoluble in the hot solvent, filter them out at the high temperature; if the impurities are soluble in the cold solvent, leave them in the solution and separate them from the solution after crystallization.

④ Form trim crystals of the purified material.

⑤ The boiling point of the solvent is neither too low nor too high. It is difficult to separate and operate when it is too low because of the small change of solubility. It is not easy to remove the traces of the solvent when its boiling point is too high.

⑥ Low price and available easily.

The common solvents used in the purification of most organic solids are: water, ethanol, acetone, petroleum ether, carbon tetrachloride, benzene and ethyl acetate, etc.

In choosing the solvent the chemist is guided by the principle "similar chemical structures will dissolve one another" or "like dissolves like" . That is the solute tends to be dissolved in a certain solvent with the analogous structure or polarity. The solubilities of some common compounds can be found in reference handbooks, such as "Solubilities of Inorganic and Organic Compounds" (Ed. Stephen, et al. , 1963). However, in practice, the suitable solvent can only be chosen from the solubility test, as shown in the following procedure:

Weigh about 0. 1g of the finely divided solid into a test tube, add 1mL of the solvent, shake and stir the mixture to see if the solid will dissolve. If not, the mixture should be warmed very gently with stirring and shaking. If the solid dissolves in the cold solvent or with gentle warming, the solvent will be unsuitable. Now carefully heat the mixture to boiling with stirring. If the solid does not dissolve readily, add more solvent in 0. 5mL portions (gently boiling and stirring after each addition) until it dissolves or until the total volume of the added solvent is about 3mL. If the solid does not dissolve (or nearly so) by this time probably, the solvent is not suitable; if it does dissolve (except for insoluble impurities), cool the solution down by scratching it to see if crystallization occurs. If crystals appear, examine them visually for apparent yield and evidence of purity (absence of extraneous color, good crystal structure). If necessary, repeat the process with other solvent until a suitable one is identified. If no single solvent is satisfactory, it is possible to employ a mixture of two solvents, termed a solvent pair. Choose one solvent in which the compound is quite soluble and the other in which it is comparatively insoluble (make sure that the two are miscible). These are solvent pairs that are often employed: the mixture of ethanol and water, or the mixture of ethanol and ethyl ether, or the mixture of ethanol and acetone, or the mixture of ethyl ether and petroleum ether, or the mixture of benzene and petroleum ether, etc.

(3) The Dissolution of the Solid

Be cautious of flammable solvents, stick to the procedure described by your supervisor.

The organic solvents are either flammable or poisonous. Maybe they are both flammable and poisonous. Remember to extinguish any flames nearby. The reaction should be carried out in the hood. A three-neck flask or a round-bottom flask is often employed as a container because its flask mouth is comparatively narrow, decreasing volatility and helping the material dissolve quickly with stirring. If the solvent has a low boiling point and is flammable, heating it directly on the asbestos gauge should be prohibited. Fit a reflux condenser on the boiling flask and select proper heat sources on the basis of the boiling point. The reflux condenser should be fitted on too if the material does not dissolve very well and its heating time is too long to protect the loss of the solvent (Figure 2.14).

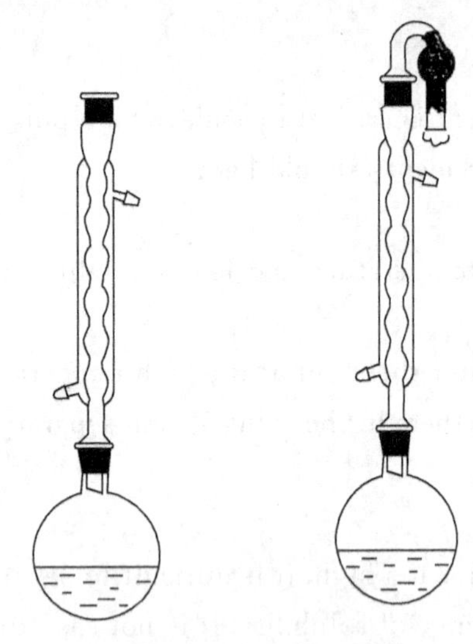

Figure 2.14 Reflux Apparatus

Put the crude product into a container with a narrow flask mouth, add some solvent less than the calculated value and then add it dropwise until the product exactly dissolves while it is heated. 20%～100% of solvent is allowed to be added to dilute the solution otherwise the crystals will separate very quickly when the hot solution is filtered. If the applied amount of the solvent is unknown, a small amount of solvent is added and the solvent is heated to the boiling point. If the solid still does not dissolve completely, add the solvent bit by bit until the solid exactly dissolves at the boiling point. This is the process of dissolving.

The product may appear in an oleaginous state in the process of dissolving. It is very disadvantageous for recrystallization because it can bear both impurities and a small amount of solvent at the same time. This phenomenon should be avoided, therefore, and it can be done in the following ways: ①The boiling point of the chosen solvent should be lower than the melting point of the solute; ②If a solvent of a lower boiling point can not be found, dissolve the material at a temperature below its melting point.

When a solvent pair is employed, the material can be dissolved in the quite soluble solvent at an elevated temperature first, and if the solution is colorized, powdered charcoal is employed to remove the colored impurities. Filter out the insoluble impurities-absorbed charcoal while the solution is hot. Heat the filtrate near to the boiling point, and add the less soluble solvent until it just reaches turbidity. If the turbidity does not vanish when it is heated, the quite soluble solvent should be added carefully until the solution becomes clear. Lay the solution in a proper place for crystallization. If the proper proportion for the recrystallization is known, prepare the solvent pair first and do the recrystallization according to the method of a single solvent.

(4) The Removal of the Impurities

Filter the hot solution to remove the insoluble impurities. Add some activated carbon to remove the colored impurities when the solution is a few degrees below the boiling point if the crude sample of a compound known to be white yields a colored recrystallization solution. See Figure 2. 3 and Figure 2. 4.

(5) The Separation of Crystals

Set aside the hot filtrate in place and cool it down slowly. It will need a few hours. Maybe it will need a longer time in some circumstances. Neither cool the filtrate down quickly because it will make the crystals very small nor make the crystals too large because it will intermix with the original liquor and make it very difficult to dry. When the large crystals are formed, stir the solution strongly and have it form small crystals.

Dropping a few seed crystals into the solution will benefit the formation of crystals or scratch the wall of the container with a glass rod if the crystals do not separate in the cold solution.

If crystals cannot be separated from some oily material, the solution should be heated until it is clear. Cool down the solution naturally and stir up acutely to disperse the oily material until the oily material disappears.

If the material cannot be purified successfully by crystallization, the other methods (such as chromatography and ion exchange resin, etc.) should be used to purify the material.

(6) The Collection and Washing of the Crystals

The crystals are collected by swirling and pouring the magma into a porcelain Büchner funnel or Hirsch funnel (small and conical shape). The funnel is fitted with a rubber stopper or conical sleeve which is seated in the neck of a heavy-wall side-arm filter flask or side-arm test tube connected to an aspirator. The apparatus for vacuum filtration of solids are showed as Figure 2. 15.

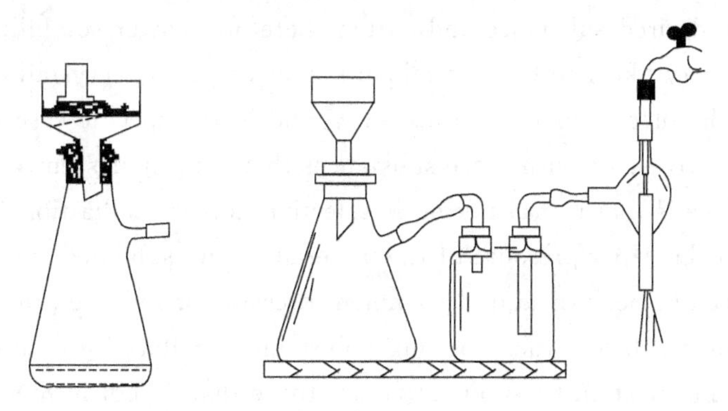

Figure 2. 15 Vacuum Filtration

These porcelain funnels are fitted with a circle of filter paper; the perforations permit liquid to drain rapidly into the evacuated flask. Be sure that the filter paper is flat. If necessary, trim the filter paper so that it covers the perforations completely but is creased into a channel at the side wall of the funnel. Wet the paper with solvent and turn on the aspirator to seat the paper firmly. After transferring the mixture to the funnel, scrape loose any liquid clinging to the wall of the flask and rinse it in with a small portion of the filtrate or a little more solvent.

The crystals should be compacted with a spatula to express as much solvent as possible. If the crystals mass is not too dense, it may be possible to remove most of the residual solvent by allowing the aspirator to pull air through the funnel for a few minutes.

The Hirsch funnel and side-arm test tube can be used in small-scale crystallizations.

(7) Drying the Crystals

The crystals must be dried by proper method as its surface adsorb a small amount of solvent after filter and washing. There are many drying method as the following: air dry; oven dry and vacuum desiccator, etc.

Notes

When you are using a water pump, it is very important to have a safety trap mounted in the vacuum line leading from the filter flask.

2.6.2 Sublimation

2.6.2.1 Principle

Sublimation is a technique that is especially suitable for the purification of solid substances at the microscale level.

It is particularly advantageous when the impurities present in the sample are nonvolatile under the conditions employed. Sublimation is a relatively straightforward method in that the impure solid need only be heated and mechanical losses are easily kept to a minimum.

The technique has additional advantages: ①it can be the technique of choice for purifying sensitive materials, as it can be carried out under very high vacuum and thus it is effective at low temperature; ②solvents are not involved and indeed, final traces of solvents are effectively removed; ③impurities most likely to be separated are those having lower vapor pressures then the desired substance and, often therefore, lower solubilities, exactly those materials that are very likely to be contaminants in attempts at recrystallization; ④solvated materials tend to dissolve during the process; ⑤and in the specific case of water as a solvate, it is very effective even with those substances that are deliquescent. The main disadvantage of the technique is that it may not be as selective as recrystallization. This lack of selectivity occurs when the vapor pressure of the materials being sublimed are similar.

Sublimation can be performed under ambient pressure or reduced pressure. Most material sublime when heated below their melting points under reduced pressure. Substances that are candidates for purification by sublimation are those that do not have strong intermolecular attractive force. Naphthalene, ferrocene and *p*-dichlorobenzene are examples of compounds that meet these requirements.

The processes of sublimation and distillation are closely related. Crystals of a solid substance that sublimes, when placed in a heated or evacuated container, will gradually generate molecules in the vapor state by the process of evaporation (i. e., the solid exhibits a vapor pressure). Occasionally, one of the vaporized molecules will strike the crystal surface or the walls of the container and be held by attractive forces. This latter process is the reverse of evaporation and is termed condensation.

Sublimation is the complete process of evaporation from the solid phase to condensation from the gas phase to form crystals directly without passing through the liquid state.

A typical single-component phase diagram is show in Figure 2.16, which relates the solid, liquid and vapor phases of a substance to temperature and pressure. Where two of the areas (solid, liquid, or vapor) touch, there is a line and along each line, the two phases exist in equilibrium. Line *BO* is the sublimation-vapor pressure curve of the substance in question, and only along line *BO* can solid and vapor exist together in equilibrium, at temperatures and pressures along the *BO* curve, the liquid state is thermodynamically unstable. Where the three phases intersect, all three phases exist together in equilibrium. This point is called the triple phase point.

Many solid substances have a sufficiently high vapor pressure near their melting point and they thus can be sublimed easily under reduced pressure in the laboratory. Sublimation occurs when the vapor pressure of the solid equals the applied pressure.

Heating the sample just below the melting point of the solid causes sublimation to occur. The vapors condense on the cold-finger surface, whereas any less volatile residue will remain at the bottom of the flask. Apparatus suitable for sublimation of small quantities are now commercially available.

A sublimation apparatus based on the entrainment principle is illustrated in following Figure 2.17. Air drawn in through the lip of the evaporating dish flows over the material to be sublimed, sweeping the vapors up through the asbestos sheet so that they condense on the funnel or glass wool. The air flow must be gentle enough that the sublimate will not pass through the glass wool and condense inside the funnel stem or rubber tubing.

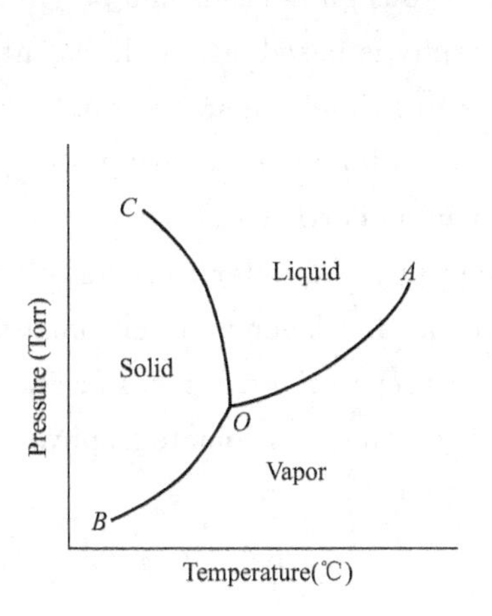

Figure 2.16 Single-component Phase Diagram

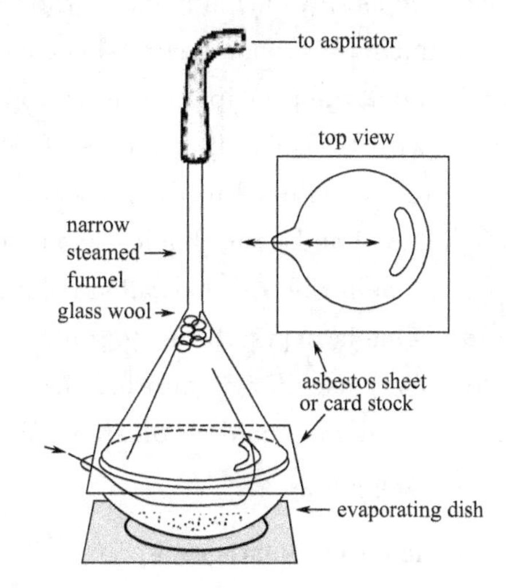

Figure 2.17 Apparatus for Sublimation

An example of the purification of a natural product is the alkaloid caffeine. This substance also can be isolated by extraction from tea.

2.6.2.2 Procedure

① Construct one of the sublimation setups described above.

② Powder the dry material to be sublimed finely (if necessary) and spread it in a thin, even layer over the bottom of the sublimation container. Assemble the apparatus and turn on the aspirator or cooling water.

③ Heat with an appropriate heat source until sublimate begin to collect on the condenser, then adjust the temperature to attain a suitable rate of sublimation without melting or charring the sublimate. If the condenser becomes overloaded with sublimate, cool and disassemble the apparatus to remove the sublimate, then reassemble it and resume heating.

④ When all of the compound has sublimed or only a nonvolatile residue remains, remove the apparatus from the heat source and let it cool.

⑤ Carefully remove the condenser and scrape the crystals into a suitable container using a flat-bladed spatula.

2.7 Chromatographic Techniques

The principle of chromatography is based on the passage of the constituents to be separated between two immiscible phase. For this, the sample is dissolved in the mobile phase, which can be either a gas or liquid phase, and moved across a stationary phase, which can either be in a column or on a solid surface.

The components of the mixture are separated after sufficient running time due to the interactions of constituents with the stationary phase.

Classification of chromatographic methods according to the mobile and stationary phase are ①gas-solid chromatography; ②liquid-solid chromatography; ③gas-liquid chromatography; and ④liquid-liquid chromatography. The solid phase used can be made of a number of material such as silica gel alumina, or ion exchange resin. The liquid are made by coating immobilized liquid on support material such as silica gel, porous glass beads and polymer. In the case of ①and ②the principle is absorption chromatography is based on the direct interaction of analyze with the solid. In the case of ③and ④, the principle is distribution. Distribution chromatography is linked to the presence of an immobilized liquid stationary phase.

Liquid-solid and liquid-liquid chromatography can be undertaken either on a column or on a plane. The former is also called column chromatography and latter consists of two common types. One is paper chromatography and the other is thin-layer plate chromatography, which is widely used for monitoring the reaction and for preparative purposes gas-solid chromatography and gas-liquid chromatography are limited to column chromatography, which is called GC commonly.

2.7.1 Column chromatography

2.7.1.1 Principle

Column chromatography is the important physical methods of the separation and purification of organic compounds, also can be used to separate a large amount of organic compounds.

In column chromatography the stationary phase is a solid, which separates the components of a liquid passing through it by selective adsorption on its surface. The types of interactions that cause adsorption are the same as those that cause attraction between molecules, that is, electrostatic attraction, complexation, hydrogen bonding, and van der Waals forces.

A column such as shown in Figure 2.18 is used to separate a mixture by column chromatography. The column is packed with a finely divided solid such as alumina or silica gel, which serves as the stationary phase, and a sample of the mixture to be separated is applied at the top. If the mixture is a solid, it must be dissolved in a solvent and then applied. The sample will initially be absorbed at the top of the column, but when an eluting solvent, the mobile phase, is allowed to flow through the column, it will carry with it the components of the mixture. Owing to the selective adsorption power of the solid phase, the components ideally will move down the column at different rates. A more weakly adsorbed compound will be eluted more rapidly than a more strongly adsorbed compound, because the former will have a higher percentage of molecules in the mobile phase. The progressive separation of the components will appear as in Figure 2.19.

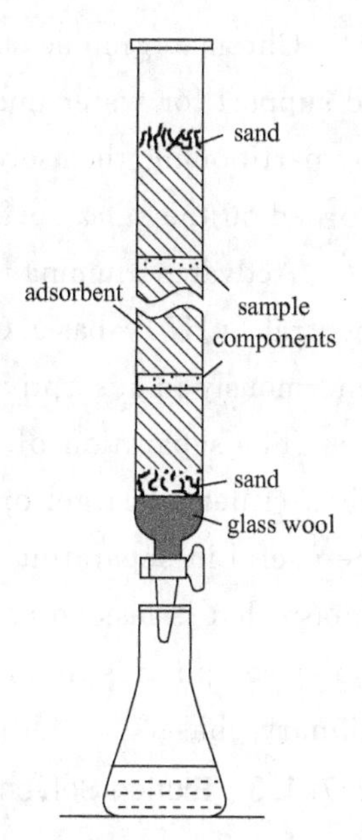

Figure 2.18 Column Chromatography

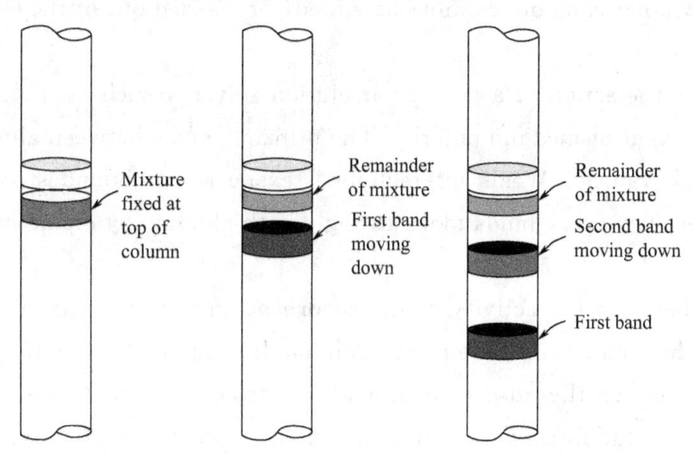

Figure 2.19 Separation of Component by Column Chromatography

2.7.1.2 Adsorbents

Silica gel ($SiO_2 \cdot xH_2O$) and aluminum oxide (Al_2O_3) are commonly used for the stationary phase in column chromatography. Both adsorbents are generally used with non-polar to moderately polar elution solvents as the mobile phase. Today most chromatographic separations use silica gel, because it allows the separation of compounds with a wide range of polarities. Silica gel also has an advantage over aluminum oxide. It causes few chemical reactions with the substances being separated. Aluminum oxide is more polar, hence polar organic compounds, such as carboxylic acids, amines, phenols, and carbohydrates, adsorb so tightly on an alumina surface that highly polar solvents are needed to dislodge them.

Chromatographic silica gel has 10%～20% adsorbed water by weight and acts as the solid support for water under the conditions of partition chromatography. Compounds separate by partitioning themselves between the elution solvent and the water that is strongly adsorbed on the silica surface.

Activated alumina is available commercially as a finely ground powder in acidic (pH4), neutral (pH7), basic (pH10) formulations and grades. Different brands and grades vary enormously in adsorptive properties.

The separation of very polar compounds may require an adsorbent of less polarity than either silica gel or alumina. For this situation, reverse-phase chromatography may be useful in separating the compounds. Reverse-phase chromatography uses a stationary phase that is less polar than the mobile phase. These conditions cause elution of the more polar compounds first, whereas the less polar compounds adsorb more tightly to the stationary phase.

2.7.1.3 Elution Solvents

In column chromatography, the solvents used to dislodge the compounds adsorbed on the column are made progressively more polar. As we discussed in the above technique on TLC, non-polar compounds bind less tightly on the solid adsorbent and dislodge easily with non-polar solvents. Polar compounds adsorb more tightly to the surfaces of metal oxides. Therefore, polar compounds must be eluted, or washed out of the column with polar solvents.

Let's compare the action of a non- polar elution solvent, such as hexane, with that of diethyl ether, a solvent of medium polarity. The primary forces between alumina and hexane molecules are weak van der Waals interactions. Hexane is not bound strongly to the column. Diethyl ether, however, binds more strongly with alumina by dipole-dipole and coordination interactions.

The balance between the activity of the adsorbent and the polarity of the solvent controls the rate at which materials elute from a column. If compounds elute too rapidly and poor separation occurs, either the adsorbent should be stronger or the solvent should be less polar. Conversely, if elution is so slow that only polar solvents are effective, a milder adsorbent is needed. The following gives the usual order of elution, from the least potent (non-polar) to the most powerful (polar) elution solvent.

Alkanes (petroleum ether, hexane, cyclohexane) $<$ Carbon tetrachloride $<$ Toluene $<$ Dichloromethane, Diethyl ether $<$ Chloroform $<$ Acetone $<$ Ethanol $<$ Methanol

2.7.1.4 Choosing a Column and Making a Macroscale Column

Once a choice about which adsorbent to use for a separation has been made, you need to decide how much adsorbent to use. This depends on the amount of material that you wish to separate, and the diameter of the column that will a short, fat one. In general, one should use 20 to 30 times as much adsorbent as the material to be separated, A 10∶1 or 15∶1 ratio of the height of the adsorbed to the column diameter is normal (Table 2.3).

Table 2.3 Representative Columns

Sample(g)	Silica gel (g)	Column dimensions diameter×height(cm)
5	150	3.5 × 40
2	60	2.0 × 30

Alumina has a bulk density of about $1g/cm^3$, but silica gel has a bulk density of about $0.3g/cm^3$. For example, 20g of alumina would fill a column having a 1.5cm diameter to a height of about 11cm. You would need to use a fatter column for the same amount of silica gel.

After you have chosen a column and weighed out the adsorbent that you need, you can prepare the column for use. The construction of such a column is just as crucial to the success of the chromatography as is the choice of adsorbent and elution solvents. If the adsorbent has cracks or channels or if it has a non-horizontal or irregular surface, you will get poor separation.

Clamp the glass chromatography tube in a vertical position onto a ring stand and fill approximately one-half of it either with the first developing solvent that you plan to use or with a less polar solvent. Cover the plug with approximately 6mm of white sand. The plug and sand serve as a support base to keep the adsorbent in the column and prevent it from clogging the tip.

Pour the adsorbent powder slowly into the solvent-filled column. The stopcock should be closed. Take care that the adsorbent falls uniformly to the bottom. The adsorbent column should be firm, but if it is packed too tightly, the flow of elution solvents will become too slow.

Gentle tapping on the side of the column as the adsorbent falls through the solvent prevents the appearance of bubbles in the adsorbent. The top of the adsorbent must always be horizontal. After all of the adsorbent has been added, carefully pour approximately 4mm of white sand on top. This layer protects the adsorbent from mechanical disturbances when new solvents are poured into the column later.

The column must never be allowed to dry out once it is made; and the solvent level should never be allowed to go below the upper level of the adsorbent.

2.7.1.5 Elution Techniques

(1) Dissolve the Sample

The mixture of compounds being separated is always poured into the column in a solution. The solvent is the preferred solvent for the sample solution. The solution should be as concentrated as possible, preferably no more than 5mL in volume. If the sample is not soluble in the first elution solvents, a very small amount of a more polar solvent can be used to dissolve them.

(2) Pour the Solution of the Sample into the Column

Before the solution of the organic mixture is applied to the column, the solvent used in making the column should be allowed to drip out until the liquid level is just at the top of the upper sand layer. Then close the stopcock and pour the mixture to be separated into the column. After reopening the stopcock and allowing the upper level of the solution to reach the top of the sand, stop the flow, fill the column with elution solvents, and proceed to develop the chromatogram. Follow the same procedure when changing elution solvents.

(3) Elute the Compounds

The less polar compounds elute first along with the less polar solvents. Polar compounds usually come out of a column only after a switch to a more polar solvent. As the development of the chromatography proceeds under the proper conditions, the compounds in the mixture put onto the top of the column will separate into distinct bands, or zones.

A mixture of two solvents is also commonly used for elution. For example, the development of the column can begin with hexane and, if nothing is eluted from the end of the column with this solvent, a 10% or 20% solution of ether in hexane can be used next. The ether/hexane solution can be followed by pure diethyl ether.

Thin-layer chromatography is one method that allows you to estimate how good a certain solvent will be as an elute for a chromatographic separation. You should use the same adsorbent material as you expect to use on the column. The range of R_f value for any of the compounds in the mixture should be about 0.2~0.8 for a useful elution solvent.

A greater height of developing solvent above the adsorbent layer will provide a faster flow through the column. An optimum flow rate is about 2mL/min. Faster rates do not allow enough time for the adsorption equilibria to take place, and incomplete separation may occur. But if the flow is too slow, poor separation may result from the natural diffusion of the bands. A reservoir at the top of a column can be used to maintain a proper height of elution solvent above the adsorbent.

The size of the elute fraction collected at the bottom of the column depends on the particular experiment. Common fraction sizes range from 25 to 150mL. If the separated compounds are colored, it is a simple matter to tell when the different fractions should be collected. However, column chromatography is not limited to colored materials. With an efficient adsorbent column, each compound in the mixture being separated will be eluted separately.

(4) Collect the Eluted Compounds

Pure components of the mixture are recovered by the evaporation of the solvent in the collected fractions. A rotary vacuum evaporator is commonly used to remove the solvents.

2.7.1.6 Summary of Column Chromatography Techniques

① Prepare a properly packed column of adsorbent.

② Add the sample mixture to column in a small volume of solution.

③ Elute the adsorbed compounds with progressively more polar elution solvents.

④ Collect the eluted compounds in fractions from the bottom of the column.

⑤ Evaporate the solvents to recover the separated compounds.

2.7.2 Thin Layer Chromatography (TLC)

2.7.2.1 Principle

Thin-layer chromatography (TLC) is one of the most widely used analytical techniques, at the same time it can be used for separating compounds at a scale between mg to g in the laboratory. TLC is a simple, inexpensive, fast, sensitive, and efficient method for determining the number of components in a mixture, for possibly establishing whether or not two compounds are identical, and for following the course of reaction.

Thin-layer chromatography involves the same principles as column chromatography, and it also is a form of solid-liquid adsorption chromatography. In this case, however, the solid adsorbent is spread as a thin layer (approximately 250μm) on a plate of glass or rigid plastic. A drop of the solution to be separated is placed near one edge of the plate, and the plate is placed in a container, called a developing chamber, with enough of the eluting solvent to come to a level just below the "spot" . The solvent migrates up the plate, carrying with it the components of the mixture at different rates. The result may then be seen as a series of spots on the plate, falling on a line perpendicular to the solvent level in the container. The retention factor (R_f) of a component can then be measured as indicated in Figure 2. 20.

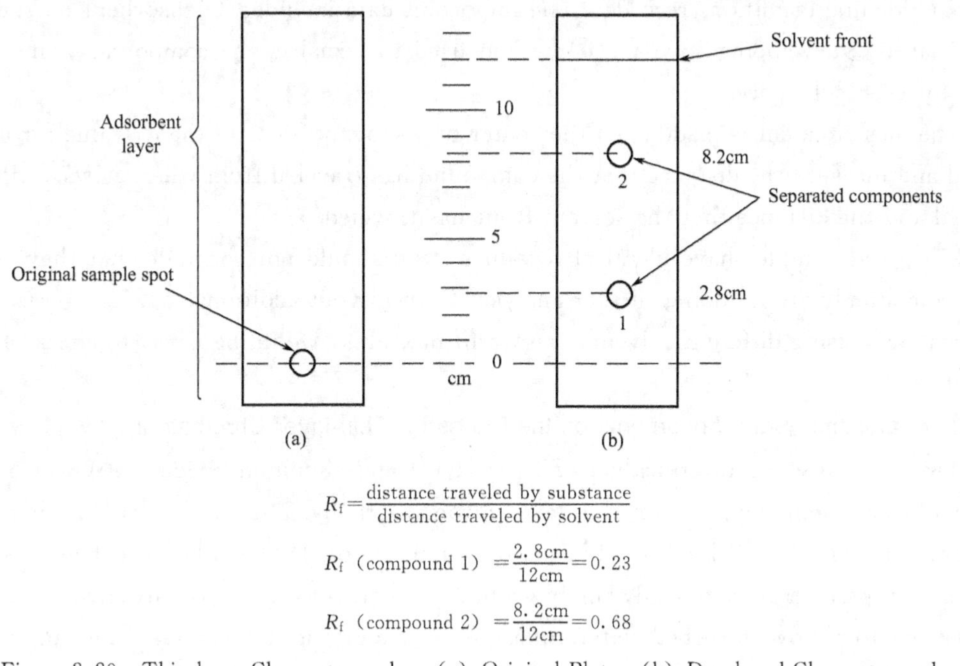

$$R_f=\frac{\text{distance traveled by substance}}{\text{distance traveled by solvent}}$$

$$R_f\ (\text{compound 1}) =\frac{2.8\text{cm}}{12\text{cm}}=0.23$$

$$R_f\ (\text{compound 2}) =\frac{8.2\text{cm}}{12\text{cm}}=0.68$$

Figure 2. 20 Thin-layer Chromatography: (a) Original Plate; (b) Developed Chromatography

This chromatographic technique is very easy and rapid to perform. It lends itself well to the routine analysis of mixture composition and may also be used to advantage in determining the best eluting solvent for subsequent column chromatography. However, it should be borne in mind that volatile compounds (b. p. <100℃) cannot analyzed by TLC.

2. 7. 2. 2 Material of Plates

Thin-layer chromatography uses glass, metal, or plastic plates coated with a thin layer of absorbent as the stationary phase.

2. 7. 2. 3 Developing and Visualization of Thin-layer Chromatography

A small amount of the mixture being separated is spotted on the absorbent near one end of the plate. Then the thin-layer plate is placed in a closed chamber, with the applied spot edge immersed in a shallow layer of developing solvent. The solvent rises through the stationary phase by capillary action. Finally the thin-layer plate is picked up from the developing chamber when the solvent front is about 1cm from the top of the plate. The position of the solvent front is marked immediately with a pencil, before the solvent evaporates. Two methods are available to visualize the spots: ①the plate can be illuminated by exposure to ultra violet radiation when it is impregnated with a fluorescent indicator; ②the other is placing the plate in a jar containing a few iodine crystals. The spots will color with the brown of iodine.

2.7.2.4 Determination of the R_f

The term R_f stands for "ratio to the front" and is expressed as a decimal fraction: R_f= Distance traveled by compound/ Distance traveled by developing solvent front. The R_f value for a compound depends on its structure and is a physical characteristic of the compound. Whenever a chromatogram is done, the R_f value is calculated for each substance and the experimental conditions recorded. The important data include: ①absorbent on the thin-layer plate; ②developing solvent; ③method used to visualize the compounds; and ④ R_f value for each substance.

The measurement is made from the center of a spot to calculate the R_f value for a compound and measure the distance that the compound has traveled from where it was originally spotted and the distance that the solvent front has traveled.

When two samples have identical R_f values, you should not conclude that they are the same compound without doing further analysis. Carrying out additional TLC analyses on the two samples, using different solvents or solvent mixtures, would be a way to check on their identities.

2.7.2.5 Common Solid Absorbents on the Plates for Thin-layer Chromatography

Two solid absorbents, silica gel ($SiO_2 \cdot xH_2O$) and aluminum oxide (Al_2O_3 also called alumina), are commonly used for thin-layer chromatography. The more polar the compound, the more strongly it will bind to silica gel or aluminum oxide. All solid absorbents used for TLC are prepared from activated, finely ground powder. Activation usually involves heating the powder to remove adsorbed water. Silica gel is acidic, and it separates acidic and neutral compounds. Acidic, basic, or neutral aluminum oxide is adapted to separate non-polar and polar organic compounds.

The plates pre-coated with a layer of adsorbent can be brought commercially. These plates are available in plastic or glass. The longer the distance that the developing solvent moves up the plate, the better will be the separation of the compounds being analyzed. Wider plates can also accommodate many more samples. Large plates with an adsorbent layer 1～2mm thick are used for preparative TLC in which samples of 50～1000mg are separated.

2.7.2.6 Sample Application

The sample must be dissolved in a volatile organic solvent. The best concentration is 1%～2%. The solvent needs a high volatility so that it evaporates almost immediately. Acetone, dichloromethane and chloroform are commonly used.

To analyze a solid, one can dissolve 20～40mg of it in 2mL of the solvent.

Tiny spots of the sample are carefully applied with a micropipet near one end of the plate. Keeping the spots small makes for the cleanest separation. Don't overload the plate with too much sample, as this leads to large tailing spots and poor separation. The micropipet is filled by dipping the constricted end into the solution to be analyzed. Hold the micropipet vertically to the plate and apply the sample by touching the micropipet gently and briefly to the plate about 1cm from the bottom edge. Mark the edge of the plate with a pencil at the same height as the center of the spot; this mark indicates the starting point of compound for your R_f calculation.

2.7.2.7 Choices of Developing Solvent

When the spots of the thin-layer plate is dry, put the thin-layer plate in a developing chamber (Figure 2.21). To ensure good chromatographic resolution, the chamber must be saturated with solvent vapors to prevent the evaporation of solvent as it rises up the thin-layer plate. Use enough developing solvent to allow a shallow layer (3～4mm) to remain on the bottom. But if the solvent level is too high, the spots may be below the solvent level. Hence the spots dissolve away into the solvent. The chromatogram will be fail.

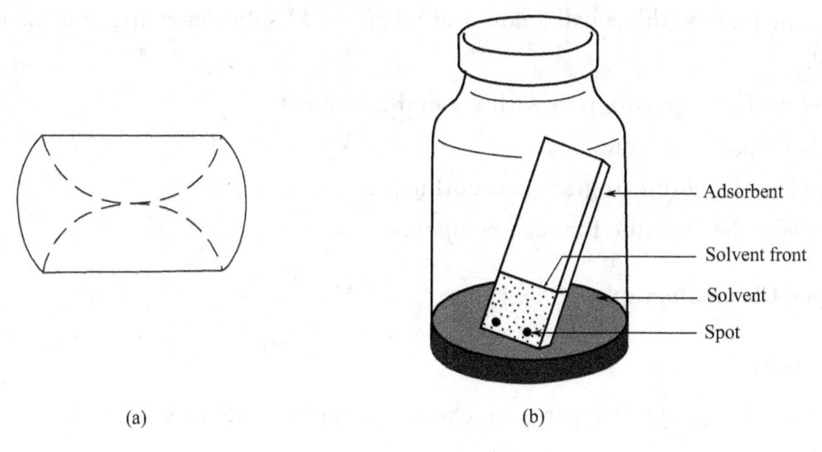

Figure 2.21 TCL Chamber

Uncap the developing chamber and place the thin-layer plate inside with a pair of tweezers. Recap the chamber, and let the solvent move up the plate. The adsorbent will become visibly moist. When the solvent front is within 0.5～1cm of the top of the adsorbent layer, remove the plate from the developing chamber with a pair of tweezers and immediately mark the adsorbent at the solvent front with a pencil. To get accurate R_f values, the final position of the front must be marked before any evaporation occurs.

2.7.2.8 Visualization Techniques

The TLC separations of colored compounds can be seen directly. However, many organic compounds are colorless, so an indirect visualization technique is needed. Commercial thin-layer plates or adsorbents that contain a fluorescent indicator are widely used. A common fluorescent indicator is calcium silicate, activated with lead and manganese. The insoluble inorganic indicator rarely interferes in any way with the chromatographic results and makes visualization straightforward. When the output from a short-wavelength ultraviolet lamp (254nm) shines on the plate in a darkened room or dark box, the plate fluoresces visible light.

Another way to visualize colorless organic compounds is by using their absorption of iodine (I_2) vapor. The thin-layer plate is put in a bath of iodine vapor prepared by placing 0.5g of iodine crystals in a capped bottle. Colored spots are gradually produced from the reaction of the separated substances with gaseous iodine. The spots are dark brown on white to tan background. After 10～15min, the plate is removed from the bottle.

Visualizing solutions containing reagents that react with the separated substances to form colored compounds can be sprayed on thin-layer plates; alternatively, the thin-layer plates can be dipped in the visualizing solution. Visualization occurs by heating the dipped or sprayed thin-layer plates with a heat gun or on a hot plate. Many of these solutions are specific for certain functional group. Two common visualizing solutions are *p*-anisaldehyde and phosphomolybdic acid.

2.7.2.9 Summary of TLC Procedure

① Obtain a pre-coated thin-layer plate of the proper size for the developing chamber.

② Spot the plate with a small amount of a 1% ～2% solution containing the materials to be separated.

③ Develop the chromatogram with a suitable solvent.

④ Mark the solvent front.

⑤ Visualize the chromatogram and outline the separated spots.

⑥ Calculate the R_f value for each compound.

2.7.3 Paper Chromatography

2.7.3.1 Principle

Paper chromatography is a partition chromatography. The filter paper is the carrier, the sample solutions are separated in the filter paper.

Paper chromatography, the water or organic solvent adsorbed on filter paper is the stationary phase. The containing water or organic solvents are mobile phase, also can be called the developing agent. It is a partition process between the two phase, it can be used for the mixtures. The separation can be due to the different distribution coefficient of the components in stationary phase and mobile phase. see Figure 2.22, Figure 2.23.

2.7.3.2 Procedure

The operation of the process of paper chromatography and thin-layer chromatography is similar. Hold the micropipet vertically to the paper and apply the sample by touching the micropipet gently and briefly to the paper where about 1cm far from the bottom edge. Mark the edge of the paper with a pencil at the same height as the center of the spot; this mark indicates the starting point of compound. A small amount of the mixture being separated is spotted on the a filter paper, hang the filter paper in a developing chamber (see Figure 2.22).

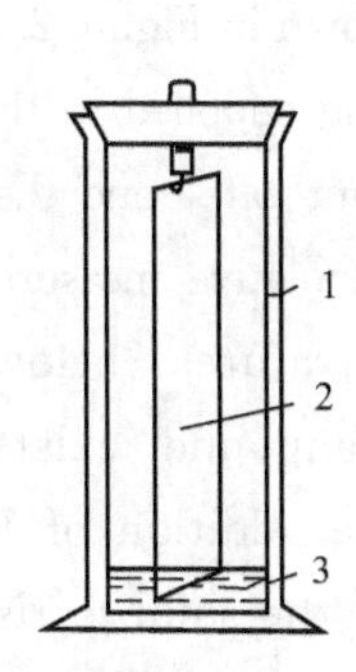

Figure 2. 22 Apparatus for Paper Chromatography
1—Paper chromatography chamber;
2—Filter paper; 3—Developed agent

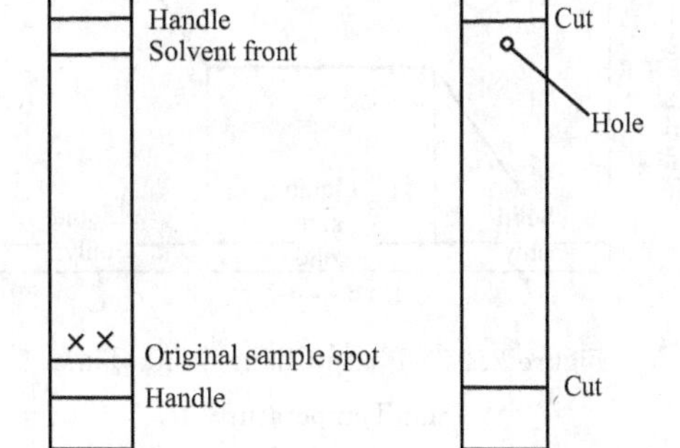

Figure 2. 23 Developed Paper Chromatography

Then the filter paper is placed in a closed chamber. When the spots of the paper is dry, with the applied spot edge immersed in a shallow layer of developing solvent. The solvent rises through the paper by capillary action. Finally the filter paper is picked up from the developing chamber when the solvent front is near the top of the paper. The position of the solvent front is marked immediately with a pencil, and then the solvent is evaporated.

The paper chromatography separations of colored compounds can be seen directly. However, many organic compounds are colorless, so an indirect visualization technique is needed. The paper can be illuminated by exposure to ultra violet radiation when it is impregnated with a fluorescent indicator or visualized reagent. (see Figure 2. 23)

Calculating R_f value of compounds, R_f value is related to the structure of separated compounds, the nature of stationary phase and mobile phase, temperature and quality of filter paper other factors. Experimental data are often different from the literature due to the effecting factors for R_f value. so you must compared with standard substance.

Advantages of paper chromatography is very easy and cheap, chromatogram can be stored for a long time; but its defect is slowly to develop.

2. 8 The Determination of the Physical Constant of Organic Compound

2. 8. 1 The Determination of Melting Point

2. 8. 1. 1 Principle

The melting point of a substance is defined as the temperature at which the liquid and solid phases exist in equilibrium with one another. If heat is added to a mixture of the solid and liquid phases of a pure substance at the melting point, ideally no rise in temperature will occur until all the solid has melted and been concerted to liquid. If heat is removed from such a mixture, the temperature will not drop until all the liquid solidifies. Thus, the melting point and freezing point of a pure substance are identical.

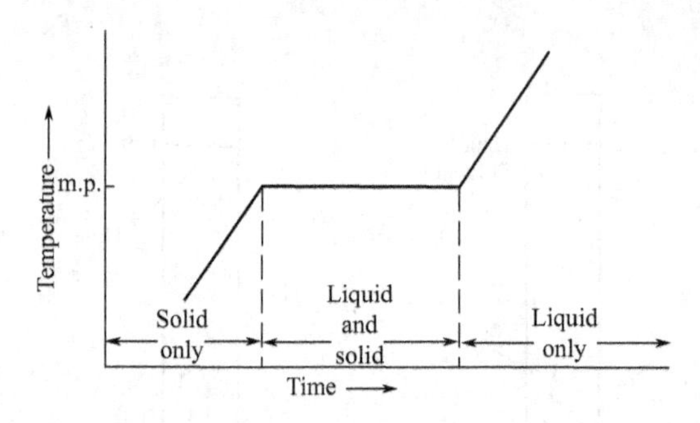

Figure 2. 24 Phase Change with Time and Temperature

The relationship among phase compositions, total supplied heat, and temperature for a pure compound is shown in Figure 2. 24. It is assumed that heat is being supplied to the compound at a slow and constant rate, and the elapsed time of heating is a cumulative measure of the supplied heat. At temperatures below the melting point, the compound exists in the solid phase, and the addition of heat causes the temperature of the solid to rise. As the melting point is reached, the first small amount of liquid appears, and equilibrium is established between the solid and liquid phases. As more heat is added, the temperature does not change, since the additional heat causes the solid to be converted to the liquid, with both phases remaining in equilibrium. When the last of the solid melts, the heat subsequently supplied causes the temperature to rise linearly at a rate that depends on the rate of heating.

2. 8. 1. 2 Melting-Point Methods and Apparatus

2. 8. 1. 2. 1 The Methods of Capillary Tube

(1) Capillary Tubes and Sample Preparation

Commercially available capillary tubes have one end already sealed and are open at the other end to permit introduction of the sample. The sample is put into the tube as follows (see also Figure 2. 25). Place a small amount of the solid whose melting point you wish to determine on a clean watch glass and tap the open end of the capillary tube into the solid on the glass so that a small amount is forced about 2～3mm into the tube. To get the solid to the closed end of the tube, take a piece of 6～8mm glass tubing about 45cm long, place this tube vertically on a hard surface (bench top or floor), and drop the capillary tube (sealed end

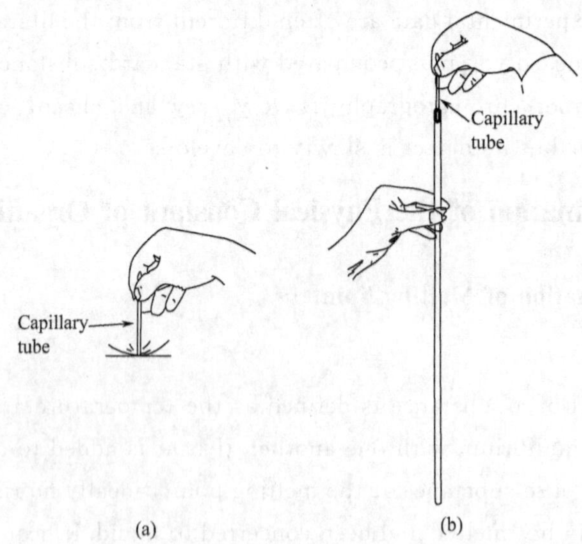

Figure 2. 25 (a) Filling a Capillary Melting Point Tube; (b) Packing Sample at Bottom of Capillary Tube

down) through the large tube several times. Surprising as it might seem, the capillary tube does not break, and the solid ends up packed at the sealed end of the capillary tube!

(2) Thiele Tube Apparatus

A simple type of melting-point apparatus is the Thiele tube, which is shown in Figure 2. 26. Tube is shaped so that heat applied to a liquid in the side-arm by a burner is distributed to all parts of the vessel by convection currents in the heating liquid so that stirring is not required. Temperature control is accomplished by adjusting the flame produced by the burner, which may seem difficult at first but can be mastered with practice.

Proper use of the Thiele tube is required to obtain reliable melting points. Secure the capillary tube to the thermometer at the position indicated in Figure 2. 26; use either a rubber band or a small segment of rubber tubing for this purpose. Be sure that the band used to hold the capillary tube on the thermometer is as close to the top of the tube as possible. Support the thermometer, with the capillary tube containing the sample already attached, in the apparatus with a cork, as shown in Figure 2. 26, or by carefully clamping the thermometer so that it is immersed in the oil. The thermometer and capillary tube must not contact the glass of the Thiele tube in any way. Make sure that the height of the heating fluid is approximately at the level indicated in Figure 2. 26. The oil will expand on heating. For this reason, the rubber band should be in the position indicated. Otherwise, the hot oil will come in contact with the rubber and melt it, and the sample tube will fall into the oil.

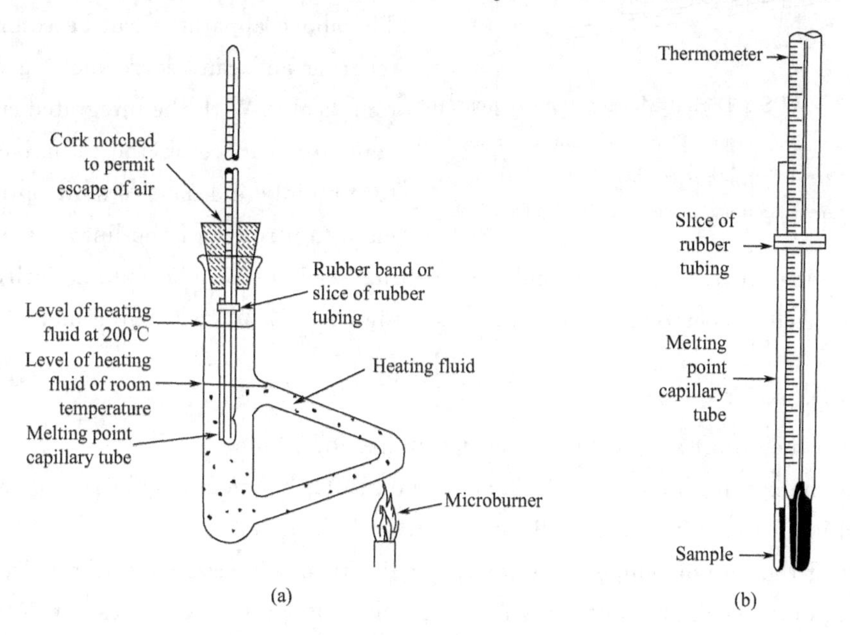

Figure 2. 26 (a) Thiele Melting-point Apparatus; (b) Arrangement of Sample and Thermometer for Determination of Melting Point

Heat the Thiele tube at the rate of one to two degrees per minute in order to determine the melting point. The maximum temperature to which the apparatus can be heated is dictated by the nature of the heating fluid.

(3) Melting-point Determination

The determination of the melting point involves taking the capillary tube containing the sample and heating it in an appropriate apparatus until the solid melts. Best results are obtained by heating the sample at the rate of about two degrees per minute. Many organic compounds undergo a change in crystalline structure just before melting, usually as a consequence of the release of solvent of crystallization. The solid takes on a softer, perhaps "wet", appearance, which may also be accompanied by a shrinkage of the sample in the capillary tube. These changes of the sample should not be interpreted as the beginning of the melting process. Wait for the first tiny drop of liquid to appear. Melting invariably occurs over a temperature range, and the melting-point range is defined as the temperature at which the first tiny drop of liquid appears up to and including the temperature at which the solid has melted completely.

2. 8. 1. 2. 2 The Method of Digital Melting-point Apparatus

(1) Digital Melting-point Apparatus

To measure the melting point, the digital melting-point apparatus is convenient, exact and easy to operate. For example, the WRS-1 digital melting-point apparatus adopts the technique such as photoelectric checking and digital temperature display, and is equipped with automatic show of early and full melt. The above apparatus can be connected to a recorder and can record melting curves automatically. With the integrated circuit, this apparatus can reach the initiative temperature rapidly and has a 6-shift optional automatic controller of the linear rate for rising or falling temperatures. The early and full melting readings can be automatically recorded without the need for personal intervention (see Figure 2. 27 WRS-1 digital melting-point apparatus).

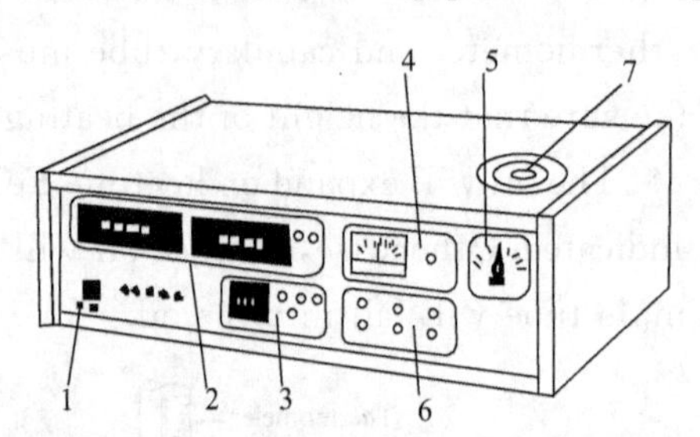

Figure 2. 27 WRS-1 Digital Melting-Point Apparatus
1—Switch; 2—Temperature; 3—Initial temperature button; 4—Adjust zero; 5—Rate selection; 6—Linearity adjust temperature and automate control; 7—Capillary jack

(2) Procedure

① Turn the on-off switch to on. Let it stabilize for 20min.

② Set the initial temperature by dialing the switch board. Press the initial temperature button and input this temperature. Then the preset light is on.

③ Select the rate of temperature for rising and turn the wave band to required place.

④ Plug in the capillary containing a sample when the preset light goes out. Now the melt light showing early goes out.

⑤ Set the ammeter to zero, press the temperature button of rising. The early melt light will remain on for several minutes, and then the full melt reading will show.

⑥ Press the early melt button, then record the early and full melt temperature until early melt reading is displayed.

⑦ Press the temperature button of falling and lower the temperature to room temperature. Turn the on-off switch to off.

2.8.2 The Determination of Boiling Point

2.8.2.1 Principle

The boiling point of a liquid is one of its important constant. it is defined as the temperature at which its vapor pressure equals the atmospheric pressure.

Every pure organic compound has a fixed boiling point at a certain pressure. Its changing range is small while a pure substance distilled, its temperature range is within 0.5～1℃. A constant thermometer reading is sometimes used as a criterion of purity of a liquid.

But sometimes an organic compound together with other components can form a binary or ternary azeotrope mixture which also has a definite boiling point. So it is difficult to say whether a liquid is a pure organic compound.

Simple distillation can be used to determinate the boiling point of a liquid organic compound. About apparatus for the determination of the boiling point is shown in Figure 2.7.

2.8.2.2 Procedure (See 2.5.1.2)

① Assemble a simple distillation apparatus.

② Put the liquid into the flask through a separatory funnel, add a couple of boiling chips, and introduce the cooling water into a condenser.

③ First heat slowly with a heating mantle, and then gradually increase the power, make the liquid boil, and start distilling.

④ Adjust the heating source, maintain distillation at a rate of 1～2drops/s, and record the temperature of first distillate (t_1).

⑤ Keep heating speed constant, continue to distill, and collect all distillates over the range of the desired temperature. Discontinue distilling any liquid when no distillates come out and the temperature decreases suddenly. Record the temperature of distillate at time (t_2).

⑥ Remove the heating source, stop the cooling water, and disassemble the apparatus following the right-to-left and top-to-bottom principle.

⑦ The boiling point range t_1～t_2.

2.8.2.3 Questions

① Why are boiling chips not added into a heated liquid?

② Why should the distillation apparatus have an opening to the atmosphere at the end?

2.8.3 The Determination of Refractive Index

2.8.3.1 Principle

Refractive or index of refraction is a physical property useful for identifying liquid or indicating their purity. Refraction index is very accurate and can be determined to four decimal places.

It will take place refraction when light from one medium to another. Then according to

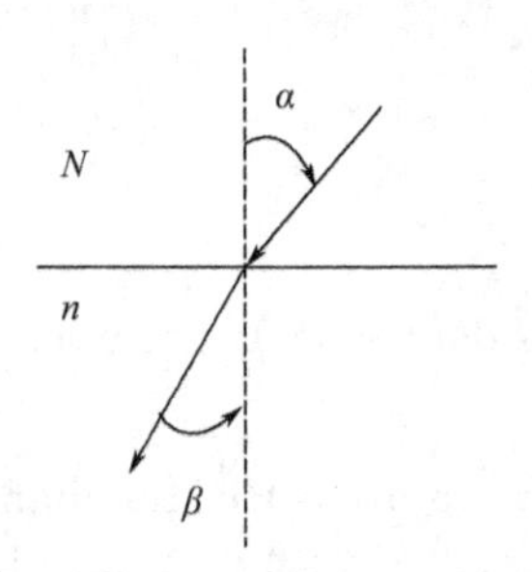

Figure 2. 28 The Refractive Phenomenon

the law of Snellius, the ratio of the sine of the angle of incidence of the beam of light striking the surface of the liquid and the angle of refraction of the beam of light in the medium is an inverse measure of two medium refractive index (N, n) for the single light of wavelength when constant condition (temperature, pressure etc.). see Figure 2. 28.

$$\frac{\sin\alpha}{\sin\beta}=\frac{n}{N}$$

If one medium is vacuum, $N=1$, then:

$$n=\frac{\sin\alpha}{\sin\beta}$$

Where in

n——the refractive index at a specified centigrade temperature and wavelength of light;

α——the angle of incidence of the beam of light striking the surface of the liquid;

β——the angle of refraction of the beam of light in the medium.

So, the refractive index n is also called absolute refractive index. Usually, n is called comparative refractive index if air is comparative standard.

The refractive index for a given medium depends on two variable factors. They are temperature and wavelength. It is usual to report refractive index indicate—measured at t℃, with a sodium discharge lamp as the source of illumination. The sodium lamp gives off yellow light of 589. 3nm wavelength, the so-called sodium D line. Under these conditions, the refractive index is shown in the following form: n_D^t. Refractive index decreases as temperature increases.

β is max when $\alpha=90°$, also called critical angle (β_0), then

$$n=\frac{1}{\sin\beta_0}$$

We can get the refractive index by determining critical angle. Usually using Abbe refractometer.

2. 8. 3. 2 The Abbe Refractometer

The instrument used to measure the refractive index is called a refractometer. The most common instrument is the Abbe refractometer. A common type of Abbe refractometer is shown in Figure 2. 29. It's popularity is due to several factors: ①the refractive index can be directly read in the same place as the adjustment image; ②only a few drops of sample are needed; ③temperature control of sample is possible; ④accuracy is very high.

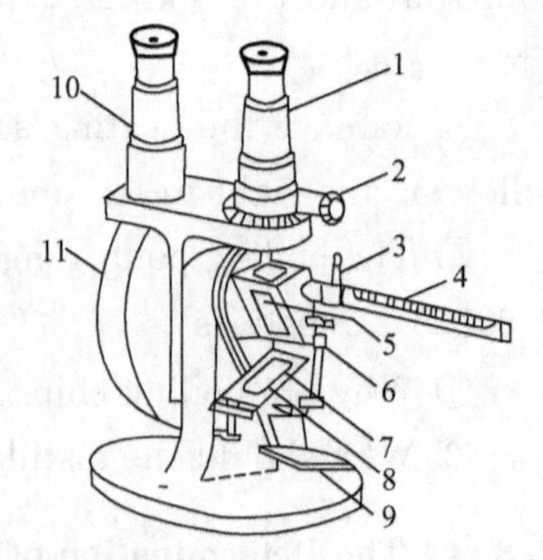

Figure 2. 29 Abbe Refractometer

1—Eyepiece with lens; 2—Chromatic-adjustment collar; 3—Water exit; 4—Themometer; 5 ~ 7—Hinged prism; 8—Groove for sample; 9—Reflector; 10—Lens to view index of refraction scale; 11—Index of refraction scale

2. 8. 3. 3 Procedure

① Begin circulation of water from the constant temperature bath well in advance of using the instrument.

② Check the surface of the prisms for residue from the previous determination. If the prisms need cleaning, place a few drops of methanol on the surfaces and blot the surfaces with lens paper.

③ Squeeze gently the prism handles and swing open the upper prism. Drop two or three drops of the liquid onto the lower prism without touching its surface[1]. Lower the upper prism and lock it into position. For volatile liquids introduce the sample from an eyedropper into the channel alongside the closed prisms.

④ Turn on the light and look into the eyepiece. Move the lamp arm and rotate the light so it shines through the window into the sample area.

⑤ Now adjust the light and the coarse adjustment knob until the field seen in the eyepieces is illuminated so that the light and dark regions are separated by as sharp a boundary as possible. Figure 2. 30 illustrates a field with dark and light portions[2].

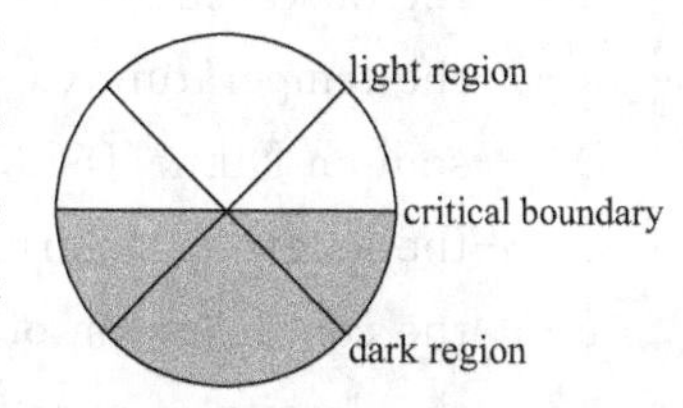

Figure 2. 30 View Through the Eyepiece When the Refractometer adjusted Correctly

⑥ Press the refractive index scale button and read the value that appears, on the field. Now move the boundary out from the cross hairs and recenter it to get a second reading take several replicate reading and report the average value. Record the temperature and the make and the model of the instrument

⑦ Taking care not to scratch the surfaces, clean the refractometer prism faces with a soft tissue paper moistened with methanol, ethanol or ether immediately after use[3]

Notes

[1] Extra special care must be taken not to scratch the surfaces of the prisms

[2] If boundary has color associated with it and/or appears somewhat diffuse, rotate the compensator drum on the face of the instrument until the boundary becomes non colored and sharp.

[3] When you determine refractive index on toxic substances work in a hood.

2. 8. 4 The Determination of the Specific Rotation

2. 8. 4. 1 Principle

The specific rotation is just as its melting point and boiling point, which a physical constant characteristic of a compound. It is important as an aid in identifying a compound or in determining its degree of purity.

The specific rotation $[\alpha]$ is the value of the optical rotation for the compound under standard condition, it can be calculated by observed rotation (α). The observed rotation (α) can be got by determination with a polarimeter.

The observed rotation is rotating degree for vibrating plane of organic compound when it is transmitted by a polarized light.

The observed rotation depends on variable factors that are the nature, concentration, reagent, temperature, wavelength and the thickness of transmitted solution. It is usual to report observed rotation indicate—measured at t℃, with a sodium discharge lamp as the source of illumination. The sodium lamp gives off yellow light of 589.3nm wavelength, the so called sodium D line. Under these conditions, the observed rotation is reported as the following form: α_D^t, The specific rotation is shown as $[\alpha]_D^t$

The specific rotation of the solution: $[\alpha]_D^t=\frac{\alpha}{Lc}$

The specific rotation of the pure liquid solution: $[\alpha]_D^t=\frac{\alpha}{Ld}$

Where in

α——the observed rotation (degree);

t——the temperature (℃);

D——sodium D line (589.3nm);

L——the length of polarimeter tube (dm);

c——the concentration of the solution (g/mL);

d——the density of pure liquid (g/mL).

2.8.4.2 The Polarimeter

Figure2.31 is Diagrammatic sketch of a polarimeter.

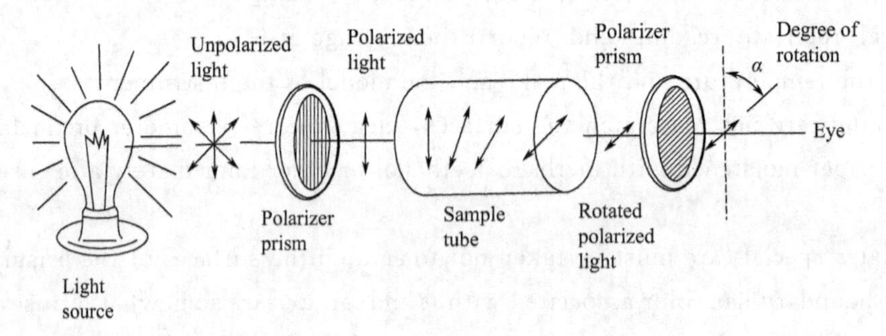

Figure 2.31 Diagrammatic Sketch of a Polarimeter

The field of vision will be formed when polarized light appears, see Figure 2.32, there are three parts.

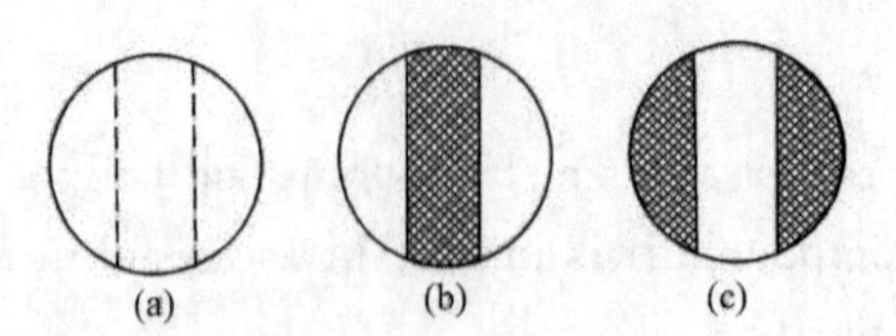

Figure 2.32 The Field of Vision

The phenomenon of the field of vision is Figure 2.32 (a), when rotating polarizer prism $\alpha=0°$, then three part is the same.

The field of vision is Figure 2.32 (b) or (c), when the solution of sample is placed to tube, then three parts are different. But the phenomenon of three parts is the same again and appears after rotating polarizer prism. The rotating degree is the observed rotation. We work out this degree by reading of scale through magnifier.

2.8.4.3 Procedure

(1) Preheating

Turn-on power, preheating 5min.

(2) Rectify of a Polarimeter

Using distilled water to rectify polarimeter.

(3) Determining the observed rotation of sample.

(4) Calculating [α] of the solution.

Chapter 3 Comprehensive Experiment

Part 1 Synthesis

3.1 Synthesis of Cyclohexene

3.1.1 Purpose

1. Learn the principle and method of preparing cyclohexene by acid-catalyzed dehydration of cyclohexanol.

2. Master the techniques of fractional distillation and simple distillation.

3.1.2 Principle

Cyclohexene is usually prepared by cyclohexanol dehydration in an acid catalyzed reaction using strong, concentrated mineral acid, such as sulfuric acid or phosphoric acid. Since sulfuric acid often causes extensive charring in this reaction, phosphoric acid, which is comparatively free of this problem, will be used.

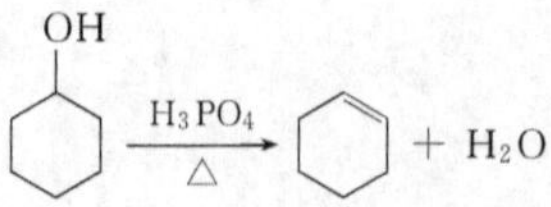

OH

$$\xrightarrow[\triangle]{H_3PO_4} \quad + H_2O$$

3.1.3 Apparatus and reagents

(1) Apparatus

Fractional distillation and distillation.

(2) Reagents

Cyclohexanol, phosphoric acid, NaCl, Na_2CO_3, $CaCl_2$ (dry).

3.1.4 Procedure

① Pour 10.4mL of cyclohexanol and 4mL of 85% phosphoric acid into a 50mL round-bottom flask. Mix the solution thoroughly by swirling the flask and add several boiling stones.

② Assemble a fractional distillation apparatus[1] as illustrated in Figure 1.7, then use a short fractional column and a 50mL flask as the receiving flask. Immerse the receiving flask up to its neck in an ice-water bath to minimize the escape of cyclohexene vapors into the laboratory.

③ Heat the reaction mixture with an asbestos pad until the products begin to boil. The temperature of the distilling vapor must not exceed 90℃[2]. The collected liquid in the receiving flask is a mixture of cyclohexene and water.

④ When only a few milliliters of residue remain in the flask, stop the distillation.

⑤ Add 1g solid sodium chloride to the distillate in the receiving flask to saturate the water layer, and shake the flask gently[3].

⑥ Decant the distillate into a separatory funnel, add 3~4mL of 5% aqueous sodium carbonate solution (or 0.5mL of 10% aqueous sodium hydroxide solution[4]) and shake the separatory funnel gently until no more gas is evolved when the funnel is vented.

⑦ Draw off the lower aqueous layer and then pour the upper layer (crude cyclohexene) through the neck of the separatory funnel into a dry small flask. Add about 1~2g of anhydrous calcium chloride to the flask and swirl the mixture occasionally until the solution becomes clear [5].

⑧ Decant the dry cyclohexene solution into a dry 50mL of round-bottom flask. Add several boiling stones and assemble the apparatus for a distillation[6]. Use a weighted 50mL round-bottomed flask submerged up to its neck in an ice-water bath as the receiver.

⑨ Collect the fraction that boils between 80~85℃ in the weighted flask, then calculate the percentage yield.

Pure cyclohexene is a colorless liquid, b. p. 82.98℃, d_D^{20} 0.8102, n_D^{20} 1.4465.

Notes

[1] The equilibrium will be shifted in favor of the product, cyclohexene, by distilling it from the reaction mixture while it is being formed.

[2] The cyclohexene (b. p. 83℃) will co-distill with the water that is also formed, (b. p. 70.8℃), and the starting material, cyclohexanol, also co-distills with the water (b. p. 97.8℃), so the distillation must be done carefully, not allowing the temperature to rise above 90℃.

[3] The salt minimizes the solubility of the organic product in the aqueous layer.

[4] As a small amount of phosphoric acid co-distills with the products, it must be removed by washing with aqueous sodium carbonate solution.

[5] The solution should become clear, and if it is still cloudy, and additional 0.5g of drying agent, swirl and last for an additional 10min.

[6] Make sure that all the glassware you will use for distilling is clean and dry. Dry all the glassware in an oven (110℃) for 10min.

3.1.5 Questions

① Why must the temperature on the top of the fractional column during the reaction period be controlled?

② What is the purpose of adding salt before the layers are neutralized and separated?

③ Why must any acid be neutralized with sodium carbonate before the final distillation of cyclohexene?

3.2 Synthesis of Ethyl Bromide

3.2.1 Purpose

Master the principle and method of preparing ethyl bromide.

3.2.2 Principle

In the laboratory ethyl bromide usually is prepared from the corresponding alcohol (ethanol) reacting sodium bromide and sulfuric acid.

Reaction equation:

Main reactions

$$NaBr + H_2SO_4 \longrightarrow HBr + NaHSO_4$$

$$C_2H_5OH + HBr \rightleftharpoons C_2H_5Br + H_2O$$

Secondary reactions

$$2C_2H_5OH \xrightarrow{H_2SO_4} C_2H_5OC_2H_5 + H_2O$$

$$C_2H_5OH \xrightarrow{H_2SO_4} CH_2{=}CH_2 + H_2O$$

$$2HBr + H_2SO_4(浓) \longrightarrow Br_2 + SO_2 + 2H_2O$$

3.2.3 Apparatus and Reagents

(1) Apparatus

Extraction/separation, fractional distillation.

(2) Reagents

Ethanol (95%), sodium bromide, concentrated sulfuric acid.

3.2.4 Procedure

① In a 100mL round-bottomed flask place 10mL of 95% ethanol and 9mL of water[1]. Cool the mixture in a cold water bath, and slowly add 19mL of concentrated sulfuric acid with thorough mixing (swirling) and cooling in the cold water bath.

② To the cold mixture add 12.9g of anhydrous sodium bromide while shaking slightly the flask to avoid formation of lumps. Introduce two boiling stones into the flask. Assemble the apparatus as shown in the Figure 1.7. Place a small amount of ice-water in the receiving flask immersed in an ice-water bath to avoid the loss of the product by evaporation[2]. The side tube of the vacuum adapter is connected with a rubber tube led to the sewer (why?).

③ Heat the flask gently over an asbestos gauge with a small flame (or heating mantle) until distillate does not contain the oily matter (about 40 minutes)[3].

④ Transfer the distillate to a separatory funnel. Draw off the organic phase (which layer?) into a dry flask immersed in an ice-water bath. To remove the impurities in the crude product, such as diethyl ether, ethanol and water, add dropwise concentrated sulfuric acid to the flask while swirling the contents until the layers have separated sharply[4,5].

⑤ Separate the sulfuric acid carefully and completely (which layer?) using a dry separatory funnel. Transfer the crude ethyl bromide to a dry distilling flask (how to transfer?), add a boiling stone and distill from a water bath using a dry apparatus.

⑥ Collect the pure ethyl bromide in a weighed bottle, which has been cooled in an ice-water bath. Collect the fraction boiling at 37～40℃, the percentage yield is 60.4%.

Pure ethyl bromide is a colorless liquid, b. p. 38. 4℃, d_4^{20} 1. 4604, n_D^{20} 1. 4239.

Notes

[1] Add a small amount of water to the reacting flask to prevent the formation of a large amount of foam during the reaction, decrease the formation of by-product diethyl ether and avoid the volatilization of hydrobromic acid.

[2] Place a small amount of ice-water in the receiving flask immersed in an ice-water bath to avoid the volatilization of the product while receiving the crude product.

[3] Control strictly the reaction temperature in order that the reaction can carry on reposefully and avoid the reactants rushing out the distilling flask due to the formation of a large amount of the foam at the beginning of the reaction. The boiling point of ethyl bromide is lower; distill slowly to avoid the loss of the product.

[4] Separate the aqueous layer as completely as possible otherwise the product will lose due to the exothermicity while washing the organic layer with concentrated sulfuric acid.

[5] The concentrated sulfuric acid is used to get rid of the impurities, such as diethyl ether, ethanol and water. When the organic layer is not washed completely, it will contain a small amount of water and ethanol. Both of them and ethyl bromide can form an azetrope respectively. Ethyl bromide and water can form an azetrope containing 1% water, the boiling point is 37℃, ethyl bromide and ethanol can form an azetrope containing 3% ethanol, the boiling point is 37℃.

3. 2. 5 Questions

① What is the main cause of the yield to be lower in this experiment?

② Which starting material is used to calculate the yield this experiment?

③ Which impurities does the crude product have? How are they removed?

④ Which measures have been take in this experiment in order to reduce the loss by the volatilization of ethyl bromide?

3. 3 Synthesis of 2-chlorobutane (*sec*-butyl chloride)

3. 3. 1 Purpose

Master the principle and method of preparing 2-chlorobutane.

3. 3. 2 Principle

Reaction equation

$$C_2H_5\underset{\displaystyle OH}{\underset{|}{C}}HCH_3 + HCl \xrightarrow{ZnCl_2} C_2H_5\underset{\displaystyle Cl}{\underset{|}{C}}HCH_3 + H_2O$$

3. 3. 3 Apparatus and Reagents

(1) Apparatus

Reflux with a gas absorption trap, extraction/separation, fractional distillation.

(2) Reagents

Anhydrous zinc chloride, *sec*-butyl alcohol, concentrated hydrochloride, sodium hydroxide solution, Anhydrous calcium chloride.

3.3.4 Procedure

① Set up a 100mL round-bottom flask with a reflux condenser connected to a gas absorption apparatus.

② Place 16g of anhydrous zinc chloride and 7.5mL of concentrated hydrochloric acid in the flask[1]. Shake the mixture until the zinc chloride is dissolved completely. Cool the solution in the flask to room temperature.

③ Add 5mL of *sec*-butyl alcohol and two boiling stones. Then boil the mixture gently over an asbestos gauge with a small flame for 40 minutes.

④ Cool the reaction mixture slightly, remove the condenser, and reassemble the apparatus for distillation. Add a boiling stone to the flask, distill and collect the fraction boiling below 115℃[2].

⑤ Transfer the distillate to a separatory funnel, separate the organic layer, and wash it with 6mL of water, 2mL of 5% aqueous sodium hydroxide solution and 6mL of water respectively.

⑥ Decant the organic layer into a small dry flask and dry it with anhydrous calcium chloride.

⑦ Decant the dried liquid into a small dry distilling flask and add two boiling stones into the flask. Distill the product from a dry fractional distillation set using a hot water bath. Collect the fraction boiling at 67～69℃ in a weighed receiver[3]. Weigh the product and calculate the percentage yield.

Pure 2-chlorobutane is a colorless and transparent liquid. b. p. 68.25℃, d_4^{20} 0.8732, n_D^{20} 1.3971.

Notes

[1] When Lucas reagent is prepared, the anhydrous zinc chloride must be dried thoroughly by fusion method. The operation of weighing the dried anhydrous zinc chloride must be quick because it is easily deliquescent.

[2] 2-Chlorobutane does not dissolve in hydrochloric acid. When the oily matter presents in the upper layer of the solution in the flask, it indicates that the reaction has occurred. The by-products, alkenes, have been removed from the product during distilling.

[3] Because the boiling point of 2-chlorobutane is lower, the operation must be quick in order to avoid the loss of the product.

3.3.5 Questions

① Why should the reflux be gentle during the reflux reaction?

② Why is the product collected by the apparatus of the fractional instead of the distillation?

③ What factors will reduce the yield of the product in this experiment?

3.4 Synthesis of diethyl ether

3.4.1 Purpose

Learn the principle and method of preparation of ethers by intermolecular dehydration.

3.4.2 Principle

Ethers are usually important organic solvents in organic synthesis as they can dissolve most organic chemicals. Normally there are two methods to prepare an ether: intermolecular dehydration of an alcohol catalyzed by concentrated acids and the Williamson synthesis from the reaction of an alkali metal salt of alcoholate with a primary alkyl halide. There are two facts that should be kept in mind: ①the dehydration method is usually performed for the preparation of a symmetric ether with a low boiling point; ②the dehydration catalyzed by concentrated acid is an equilibrium and some measures should be taken to shift the equilibrium reaction to right, such as increasing the amount of one of the reactants or removal of the products from the reaction mixture. In the present experiment, concentrated acid is employed to achieve the intermolecular dehydration of ethyl alcohol and formation diethyl ether while water, one of the products, can be removed from the reaction mixture by water separator. The reaction is shown below:

$$2CH_3CH_2OH \xrightleftharpoons[140^\circ C]{H_2SO_4} CH_3CH_2OCH_2CH_3 + H_2O$$

The pure diethyl ether is colorless liquid, its b. p. is 34.5℃, $n_D^{20}=1.3526$, it is insoluble for water, but it is dissolved by organic regent, it is very volatilizable.

3.4.3 Apparatus and Reagents

(1) Apparatus

Three-neck round-bottom flask (150mL), addition funnel (150mL), round-bottom flask (150mL), still head, thermometer (200℃), vacuum distillation adapter, condenser, separatory funnel, graduated cylinder (25mL).

(2) Reagents

Ethanol (95%), H_2SO_4 (18.4mol/L), NaOH (1mol/L), saturated aqueous NaCl, saturated aqueous $CaCl_2$, anhydrous $CaCl_2$.

3.4.4 Procedure

3.4.4.1 Preparation

① Place 25mL of ethanol (95%) and about 25mL of concentrated sulfuric acid into a 150mL three-neck flask. Shake or swirl and mix them fully, and add a boiling stone into the flask.

② Set up the apparatus according to Figure 3.1, a thermometer is placed in one neck

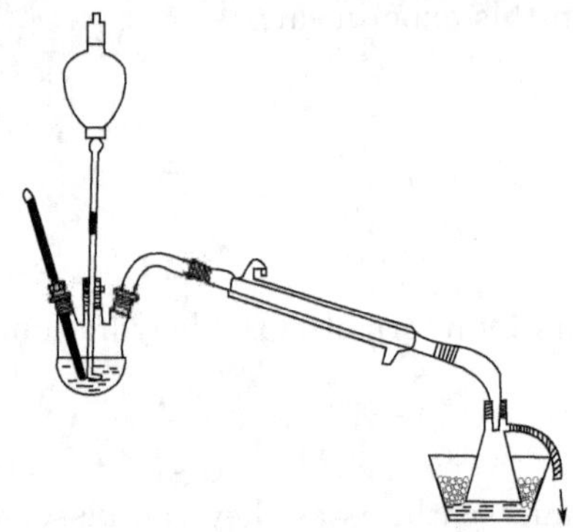
Figure 3.1 Apparatus for the synthesis of diethyl ether

while a apparatus of distillation in side neck. addition funnel of contained 50mL of ethanol is placed in the mid of neck. Make sure that all connection is tight.

③ Carefully heat the flask until the temperature of the content increases to 140℃. Then dropping ethanol, to maintain dropping and distillation at a rate of 1drop/s, keep the range 135～145℃. Continue to heat 5～10min until no distillate when ethanol is dropped.

3.4.4.2 The purify

① Pour out the fraction into separatory funnel, wash and extract with the same volume NaOH (1mol/L), Shake and swirl the funnel well, then let it stand for a moment to make sure layers well separated. Discard the lower the aqueous NaOH, wash it with 15mL of saturated aqueous NaCl, Discard the lower the saturated aqueous NaCl; finally, wash it twice with the same volume of saturated aqueous $CaCl_2$. Shake and swirl moment, again give out saturated aqueous $CaCl_2$. Separate the upper layer as product, diethyl ether, pour it into flask, dry it with 2～3g of anhydrous $CaCl_2$ until it is clarity.

② Heating and distilling the diethyl ether on the water bath of 50～60℃ after drying, collect the fraction of 33～38℃.

3.4.5 Questions

① How do you strictly control the temperature of the reaction and how do you know the reaction is nearly completed?

② What is the purpose of each washing step?

3.5 Synthesis of Phenetole (ethyl phenolate)

3.5.1 Purpose

Master the principle and method of preparing phenetole.

3.5.2 Principle

Reaction equation

$$PhOH + NaOH \longrightarrow PhONa + H_2O$$

$$PhONa + CH_3CH_2Br \longrightarrow PhOCH_2CH_3 + NaBr$$

3.5.3 Apparatus and Reagents

(1) Apparatus

Three-neck flask, mechanical stirrer, pressure-equalizing dropping funnel, reflux, extraction /separation, distillation apparatus.

(2) Reagents

Phenol, sodium hydroxide, ethyl bromide, diethyl ether, sodium hydroxide solution, anhydrous calcium chloride.

3.5.4 Procedure

① Set up a 100mL three-neck flask with a pressure-equalizing dropping funnel, reflux condenser and mechanical stirrer.

② Add 3.75g of phenol, 2g of sodium hydroxide and 2mL of water to the flask. Stir the mixture until the solids dissolve completely.

③ Heat the flask in an oil bath and control the temperature of the oil bath between 80～90℃. From the dropping funnel, add 4.25mL of ethyl bromide dropwise (require about 40min)[1]. After the addition is complete, maintain the temperature while stirring for 1 h, cool the mixture to room temperature.

④ Add right amount water (about 5～10mL) to the flask to dissolve the produced solid completely.

⑤ And transfer the solution to a separatory funnel. Separate and save the organic layer, then extract the aqueous layer with 4mL of diethyl ether. Combine the extract with the reserved organic layer, wash it twice with a same volume of saturated sodium chloride solution. Separate the organic layer, then extract the aqueous layer with 3mL of diethyl ether each time.

⑥ Combine the organic layer with the extracts and dry it with anhydrous calcium chloride. Decant the dried liquid into a dry round-bottom flask of proper size and distill off the ether from a water bath[2]. Continue to distill the residue at atmospheric pressure. Collect the product boiling at 168～171℃ in a weighed bottle.

Pure phenetole is a colorless and transparent liquid, b. p. 170℃, d_4^{15} 0.9666, n_D^{20} 1.5076.

Notes

[1] Because of the low boiling point of ethyl bromide, the flow of the condensate water must be big during the reflux to insure the enough ethyl bromide to participate the reaction.

[2] Distilling diethyl ether, do not use the open flame to heat. The tail gas should be led to the sewer to prevent the fire by the leak of the diethyl ether vapor.

3.5.5 Questions

① Preparing phenetole, what is purpose of washing the organic phase containing the product with saturated sodium chloride solution?

② What is the reflux liquid in the process of the reaction? What is the forming solid? Why the reflux is not obvious in the reaction?

3.6 Synthesis of Benzoic Acid and Benzyl Alcohol

3.6.1 Purpose

① Learn the principle and method of preparing benzoic acid and benzyl alcohol from benzaldehyde through the Cannizzaro reaction.

② Master the techniques of distillation and recrystallization.

3.6.2 Principle

When an aldehyde with no *α*- hydrogen reacts with concentrated alkali solution, there occurs self oxidation-reduction between molecules, in which one molecule of aldehyde is reduced to an alcohol and another is oxidized to an acid. This kind of reaction is called the Cannizzaro reaction.

$$C_6H_5CHO + NaOH \longrightarrow C_6H_5COONa + C_6H_5CH_2OH$$

$$C_6H_5COONa \xrightarrow{H^+} C_6H_5COOH$$

3.6.3 Apparatus and reagents

(1) Apparatus

Flask, separatory funnel, round-bottom flask, condenser, thermometer, hot funnel.

(2) Reagents

Benzaldehyde, NaOH, HCl, $MgSO_4$ (dry), saturated aqueous $NaHSO_3$, Na_2CO_3, diethyl ether.

3.6.4 Procedure

① Dissolve 8.8g of solid sodium hydroxide[1] in 8.8mL of water by swirling it in a 100mL of flask. Cool the mixture down to room temperature.

② Place 10.3mL of fresh-distillated benzaldehyde into the NaOH solution batch-wise. Cork the flask securely and shake the flask vigorously until a wax-like solution is formed[2].

③ Allow the stoppered flask to stand overnight or until the next laboratory period.

④ Add about 30～50mL of water to the mixture, stopper the flask, and shake it.

⑤ If some crystals originally present do not dissolve, add a little more water, and break up the solid with a glass rod, then shake the mixture. Repeat the procedure until all solids are in the solution.

⑥ Pour the solution into a separatory funnel, and extract the solution three times with 8mL of ether each. Combine the ether solutions, wash it with 4mL of saturated aqueous sodium bisulfite, 8mL of 10% aqueous sodium carbonate and 8mL of water, respectively[3].

⑦ Dry the organic solution with anhydrous magnesium sulfate.

⑧ Heat the crude product in a water bath, distill the ether fraction out[4]. Then drain the water out of the condenser of the apparatus, continue to distill benzyl alcohol and collect the fraction that boils between 198～204℃.

⑨ Weigh the product and calculate the yield. Pure benzyl alcohol is a colorless liquid, b. p. 205. 3℃, d_D^{20} 1. 0419, n_D^{20} 1. 5396.

⑩ Pour the alkaline aqueous solution from the separatory funnel into a mixture of 32mL of concentrated hydrochloric acid, 32mL of water, and about 20g of crushed ice. Stir vigorously during the addition.

⑪ Cool the resultant mixture in an ice bath, and collect the solid benzoic acid by vacuum filtration.

⑫ Recrystallize the crude benzoic acid from water, dry the product, and determine its melting point.

⑬ Weigh the product and calculate the yield. Pure benzoic acid is a white needle shaped crystal, m. p. 122. 13℃.

Notes

[1] Sodium hydroxide solution is corrosive, avoid splashing it on your skin. If you do so, wash it immediately with cold water.

[2] Crystallization should occur in the meantime.

[3] Ether is flammable. It should be far away from the fire.

[4] Remove the benzaldehyde by formation of its sodium bisulfite addition product.

3. 6. 5 Questions

① The Cannizzaro reaction occurs much more slowly in dilute sodium hydroxide solution than in concentrated solution. Why?

② By what means would aqueous sodium bisulfite remove unchanged benzaldehyde from the reaction mixture?

3. 7 Synthesis of Acetophenone

3. 7. 1 Purpose

① Learn the principle and method of preparing acetophenone by the Friedel-Crafts acylation.

② Master the purification technique of normal distillation for liquid with a high boiling point.

3. 7. 2 Principle

The most important method for preparing acetophenone is Friedel-Crafts acylation. The acylation of benzene by an acid chloride or anhydride is a good synthetic reaction as the reaction proceeds readily and the electron-withdrawing effect (deactivation of the aromatic nucleus) of the carbonyl group greatly decreases the likelihood of the production of di-or polyacyl derivatives. In today's experiment we shall prepare a sample of acetophenone (phenyl methyl ketone) using acetic anhydride, rather than more reactive acetyl chloride, as the acylating agent.

The benzene involved must be well dried with calcium chloride. The acetic an hydride should have a boiling point between 137～140℃. Benzene acts as both the reactant and the solvent, so its amount is excessive. Aluminum trichloride is used as the catalyst, and since it is a Lewis acid and reacts with the product to form a stable complex, its amount is also greater than the stoichiometric amount.

$$C_6H_6 + (CH_3CO)_2O \xrightarrow{AlCl_3} C_6H_5COCH_3 + CH_3COOH$$

3.7.3 Apparatus and reagents

(1) Apparatus

Mechanical stirrer, isobaric dropping funnel, condenser, vacuum pump, three-neck flask, thermometer, separatory funnel.

(2) Reagents

Benzene, acetic anhydride, anhydrous aluminum chloride, HCl.

3.7.4 procedure

① A 100mL three-neck flask is equipped with a 10mL isobaric addition funnel, a mechanical stirrer and a refluxing condenser[1]. On the top of the condenser, a drying tube with calcium chloride as drier is fitted and connected with a rubber tube to a gas trap for hydrogen chloride adsorption (see Figure 3.2).

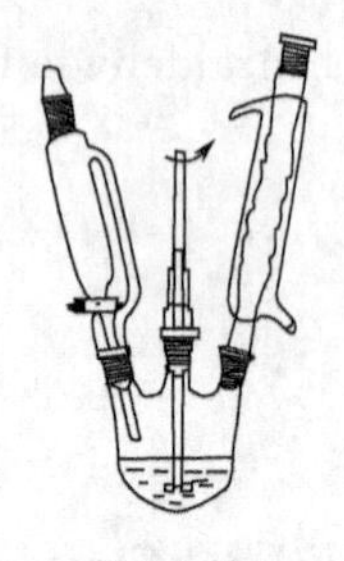

Figure 3.2 Apparatus for Synthesis of Acetophenone

② Check the apparatus and make sure the mechanical stirrer works well. Take off the dropping funnel, place 10g of anhydrous aluminum chloride[2] and 16mL of anhydrous benzene[3] quickly, and fit again the condenser and close the stopcock immediately.

③ Place 4mL acetic anhydride[4] into the dropping funnel and a stopper on the top. Turn on to stir the reaction mixture well. Then acetic anhydride[4] is added into the mixture dropwise. Be careful to control the addition rate or cool the reaction flask with water occasionally in order to avoid the reaction mixture boiling dramatically (the process needs about 10 minutes).

④ After all the acetic anhydride has been added and the vigorous reaction becomes calm, heat the reaction flask (while stirring) in a water bath (about 95℃) until no hydrogen chloride evolves (the process may need 50 minutes).

⑤ Move the water bath away and cool the flask down. Pour the product into a mixture of 18mL concentrated hydrochloric acid[5] and 30～40g crushed ice with stirring by a glass rod. Add some more hydrochloric acid to the mixture if some solids still exist.

⑥ The mixture is transferred into a separatory funnel and the upper oil layer is transferred into an flask. The aqueous solution is extracted twice with diethyl ether (8～10mL each time), then combine the extracted ether solution with the oil layer in the flask. Wash the combined solution with 8～9mL aqueous NaOH solution (3mol/L) and 9mL water successively in the separatory funnel. Then dry the organic layer with 2.5g anhydrous magnesium sulfate.

⑦ The dried diethyl ether solution is filtered into a 100mL round-bottom flask, and the distilling apparatus is assembled. Heat the crude product in a water bath first and recover benzene and diethyl ether. Then heat it on the asbestos pad. Distill and collect the fraction of 195～202℃. Weigh the product and calculate the yield.

⑧ About 2.5～3.0g (yield 50%～60%) product can be obtained. acetophenone is a colorless liquid, b.p. 202℃, n_D^{20} 1.5372.

Notes

[1] The chemicals and apparatus employed in this experiment must be dried adequately except the hydrogen chloride absorbing apparatus.

If you want to simplify the apparatus, a two-neck boiling flask with a magnetic should also be used successfully in this reaction, instead of the mechanical stirrer.

[2] The aluminum chloride used in the experiment should be in small lumps or in the form of a coarse powder, meanwhile should fume in moist air, and make a hissing sound upon the addition of a little water.

The operation of weighing and adding aluminum must be quick. Fix the cover of the container once aluminum chloride has been taken out.

[3] Benzene can be used only after it has been dried by anhydrous calcium chloride overnight or newly dried with sodium wire.

[4] Acetic anhydride must be redistilled before use and the fraction of 137～140℃ can be employed.

[5] When acid is added to the reaction mixture, a decomposition reaction of acetophenone-aluminum chloride complex takes place and releases the free acetophenone. The reaction equation is:

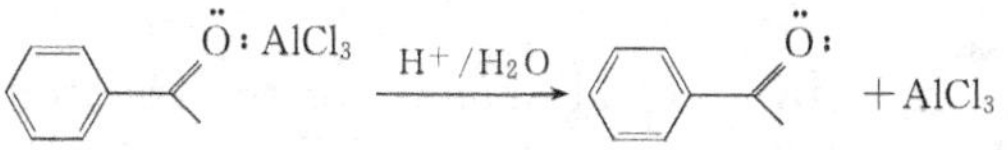

$$C_6H_5C(CH_3)=\ddot{O}:AlCl_3 \xrightarrow{H^+/H_2O} C_6H_5C(CH_3)=\ddot{O}: + AlCl_3$$

3.7.5 Questions

① In the gas trap for hydrogen chloride, please explain why the outlet of the hydrogen chloride should be far away from the water surface or be dipped into water.

② What are the key factors to accomplish this reaction and obtain a high yield of the product? What effects can be anticipated if the apparatus or reactants contain a small amount of water?

3.8 Synthesis of Benzophenone (Friedel-Crafts acylation method)

3.8.1 Purpose

Master and know the principle and method of preparing benzophenone.

3.8.2 Principle

Benzophenone is a colorless and lustrous crystal with the sweet taste and rose fragrance, thus it can be used as the spice and endow with the sweet taste to the essence. It is applied to make perfumes, fancy soaps and essences. It can be used as a starting material for the manufacture of organic dye and insecticides, as well as the production of benztropine hydrobromide and diphenhydramine hydrochloride in pharmaceutical industry.

There are many methods for preparing benzophenone. It can be prepared by either benzyl chloride as starting material via the reaction of alkylation, oxidation etc., or by benzene as starting material via the reactions of alkylation, hydrolysis, etc. In this experiment, benzophenone is synthesized by the acylation reaction using benzoyl chloride and benzene as the starting materials.

Reaction equation

$$C_6H_5COCl + C_6H_6 \xrightarrow{AlCl_3} C_6H_5COC_6H_5 + HCl$$

3.8.3 Apparatus and Reagents

(1) Apparatus

Three-neck flask, mechanical stirrer, pressure-equalizing dropping funnel, reflux, extraction / separation, distillation.

(2) Reagents

Anhydrous aluminum chloride, anhydrous benzene, benzoyl chloride, 5% sodium hydroxide solution, hydrogen chloride, anhydrous magnesium sulfate.

3.8.4 Procedure

① Set up a dry 100mL three-neck flask with a mechanical stirrer, a reflux condenser and a Claisen adapter.

② Place a thermometer in the central opening of the adapter and a pressure-equalizing dropping funnel in the side arm. Attach a drying tube filled with calcium chloride at the top of the condenser and connect the open end of the drying tube to a gas absorption trap to absorb hydrogen chloride that evolves during the reaction. The trap is made up of an inverted funnel in the beaker. Place the 5% aqueous sodium hydroxide solution used as the absorbing agent in the beaker. The inverted funnel should just touch the surface of the liquid in the beaker (the distance between the lower edge of the funnel and the liquid surface is about 1～2mm), not be immersed in the liquid to prevent the back flow of the liquid in the beaker.

③ Place 3.75g of anhydrous aluminum chloride and 15mL of dried benzene into the flask quickly, add dropwise 3mL of from the pressure-equalizing dropping funnel to the flask while stirring at room temperature. Control the dropping rate to maintain the temperature of the reaction at 40℃.

④ The mixture in the flask reacts intensely while producing hydrogen chloride and the reaction solution becomes brown gradually. After the addition, heat the solution in a hot water bath at 60℃ while stirring until no hydrogen chloride is evolved from the flask. This process requires about 1.5h.

⑤ Allow the solution to cool, and then pour the contents in the flask into a beaker containing 25mL of ice water in the hood. The precipitate separates from the solution at the same time.

⑥ Add dropwise 1~2mL of concentrated hydrochloride acid into the beaker while stirring until the precipitate is decomposed completely.

⑦ Transfer the solution to a separatory funnel, separate the organic phase, and extract the aqueous phase with 2 ×10mL portions of benzene.

⑧ Combine the extracts with the organic phase. Wash the organic phase with 10mL of water, then with 10mL of 5% sodium hydroxide solution, and finally with water three times (10mL of water each time) until the organic phase is neutral.

⑨ Dry it over anhydrous magnesium sulfate. Remove the solvent by distillation to obtain the crude product.

⑩ Distill the residue from a small distillation set under reduced pressure, collecting the fraction boiling at 187~189℃/2.00Pa (15mmHg). The product is colorless and transparent liquid. After it is cooled in a refrigeratory, it solidifies and the white crystals are obtained.

The crude product is also purified by recrystallization from petroleum ether (b. p. 60~90℃) instead of the vacuum distillation. After drying, weigh it, determine the melting point and the infrared spectrum, calculate the yield.

Pure benzophenone is a colorless crystal, m. p. 47~48℃, b. p. 305.4℃.

Notes

Benzophenone has many kinds of crystal forms. The melting points of them are different. It is 49℃ in α type, 26℃ in β type, 45~48℃ in γ type, 51℃ in δ type. Among them, the α type is more stable.

3.8.5 Questions

① Whether the polyacylated aromatic hydrocarbon is formed easily in the acylation reaction?

② Comparing with aliphatic ketone, which direction will the infrared absorption peak of carbonyl group in arone move toward (high wave number or low wave number)? Why?

③ Why can nitrobenzene be used as the solvent in Friedel-Crafts reaction? Why is it unfavourable to the Friedel-Crafts reaction that a substituent group, such as OH、OR, or NH_2, etc., exists in the aromatic ring?

④ Why should the heating bath be used in vacuum distillation, rather than heat the flask over an asbestos gauge directly, even if the boiling point of the compound is high?

3.9 Synthesis of Cyclohexanone

3.9.1 Purpose

Master the principle and the method of preparing cyclohexanone by oxidation.

3.9.2 Principle

Reaction

$$\text{Cyclohexanol (}C_6H_{11}OH\text{)} \xrightarrow[H_2SO_4]{Na_2Cr_2O_7} \text{Cyclohexanone (}C_6H_{10}=O\text{)}$$

3.9.3 Apparatus and Reagents

(1) Apparatus

Refux, extraction/separation, distillation.

(2) Reagents

Cyclohexanol, sodium dichromate, sodium chloride concentrated sulfuric acid, anhydrous magnesium sulfate.

3.9.4 Procedure

① In a 100mL beaker, dissolve 5.25g of sodium dichromate dihydrate in 30mL of water. Add carefully 4.3mL of concentrated sulfuric acid with stirring, and cool the deep orangered solution to 30℃。

② Place 5.00g of cyclohexanol in a 100mL two-neck flask and add the dichromate solution in one portion to the flask. Swirl the mixture to insure thorough mixing and observe its temperature with a thermometer.

③ The mixture rapidly becomes warm. When the temperature reaches 55℃, cool the flask in a cold water bath, and regulate the amount of cooling so that the temperature remains between 55℃ and 60℃. The temperature falls after 0.5h. Remove the cold water bath, allow the flask to stand for l h, occasional shaking it. The reaction solution becomes greenish black

④ Add 30mL of water and two boiling stones to the flask and assemble an apparatus for distillation[1].

⑤ Distill the mixture until about 25mL of distillate, consisting of water and an upper layer of cyclohexanone, has been collected (cyclohexanone and water can form an azeotrope, the boiling point is 95℃).

⑥ Saturate the aqueous layer with sodium chloride, separate the organic layer[2].

⑦ And extract the aqueous layer twice with 2×15mL portions of diethyl ether. Combine the solvent extracts with the cyclohexanone layer and dry it over anhydrous magnesium sulfate.

⑧ Filter the dried solution into a distilling flask of suitable size and carefully distill the ether from a water bath. Remove the water-cooled condenser and attach an air-cooled condenser. Distill the residual cyclohexanone at atmospheric pressure and collect the fraction boiling at 151～155℃. The yield is 3.0～3.50g, the percentage yield is 61%～71%.

Pure cyclohexanone is a colorless liquid, b.p. 155.7℃, d_4^{20} 0.9478, n_D^{20} 1.4507.

Notes

[1] It is substantively a simple steam distillation for distilling the mixture of the product

and water. Cyclohexanone and water can form an azeotrope containing cyclohexanone 38.4%, the boiling point is 95℃.

[2] The distillation yield of the water should not be too much. Otherwise a small amount of cyclohexanone will dissolve in water and result in the loss of it, in spite of using the salting out. The solubility of cyclohexanone in water is 2.4g/100mL at 31℃. Add sodium chloride to the distillate in order to decrease the solubility of cyclohexanone in water and be favorable to separate the layers.

3.9.5 Questions

① Why should the mixture of sodium dichromate and concentrated sulfuric acid be cooled below 30℃ before use?

② What is the action of the salting-out?

③ Whether the potassium permanganate can be used to oxidize the cyclohexanol under condition of the alkaline? What product will be obtained?

3.10 Synthesis of Adipic Acid

3.10.1 Purpose

① To know the principle of preparing adipic acid,

② Master the different method of preparing adipic acid.

3.10.2 Principle

Adipic acid is one of the main raw materials for manufacture of nylon 66. It can be prepared either from cyclohexanone via oxidation reaction using potassium permanganate as an oxidizer or from cyclohexanol via oxidation reaction using nitric acid or potassium permanganate as an oxidizer.

Reaction equation

Method 1

$$\text{(cyclohexanone)} \xrightarrow{KMnO_4} HOOC(CH_2)_4COOH$$

Method 2

$$\text{(cyclohexanol)} \xrightarrow{[O]} \text{(cyclohexanone)} \xrightarrow{[O]} HOOC(CH_2)_4COOH$$

3.10.3 Apparatus and Reagents

(1) Apparatus

Mechanical stirrer, reflux, suction filtration, recrystallization apparatus.

(2) Reagents

Method 1

Cyclohexanone, potassium permanganate, sodium hydroxide, sodium bisulfite, concentrated hydrochloric acid.

Method 2

Cyclohexanol, potassium permanganate, sodium carbonate, concentrated sulfuric acid.

3.10.4 Procedure

Method 1

① Set up a 100mL three-neck flask with a mechanical stirrer, thermometer and reflux condenser.

② Place 6.3g of potassium permanganate, 50mL of 0.3mol/L sodium hydroxide solution and 2mL of cyclohexanone in the flask. Note: If the reaction temperature rises above 45℃ by itself, should cool appropriately the flask in a cold water bath. Keep the temperature at approximately 45℃ for 25 minutes. Then heat the flask over an asbestos gauge until the mixture boils for 5 minutes to complete the oxidation and coagulate the manganese dioxide.

③ Test for residual potassium permanganate by placing a drop of the reaction mixture from the tip of a stirring rod on a piece of filter paper. Unreacted potassium permanganate will appear as a purple ring around the brown manganese dioxide. If unreacted permanganate remains, decompose it by adding small portions of solid sodium bisulfite until the spot test is negative.

④ Suction-filter the mixture, thoroughly wash the manganese dioxide on the filter with water.

⑤ Concentrate the filtrate to about 10mL by transferring it to a beaker, adding a boiling stone, and boiling on a hot plate or over a flame. Acidify the solution with concentrated hydrochloric acid, first by adding acid to pH 1 to 2 (pH paper) and then an additional 2mL of concentrated hydrochloric acid.

⑥ Allow the solution to cool to room temperature and collect the crystallized adipic acid by suction filtration.

⑦ The crude adipic acid can be recrystallized from water and decolorized by charcoal. The white crystals are obtained. The yield is 1.5g, the percentage is 53%, the melting point is 151～152℃.

The pure adipic acid is a white prismatic crystal, m. p. 153℃.

Method 2

① Fit a 250mL three-neck flask with a mechanical stirrer and a thermometer.

② Add 2.6mL of cyclohexanol and sodium carbonate solution (dissolve 3.8g of sodium carbonate in 35mL of warm water)[1]. Add 12g of powdered potassium permanganate in batches to the flask while stirring and controlling the temperature above 30℃[2]. The whole process needs about 2.5h.

③ After adding, continuously stir the mixture until the reaction temperature no longer rises.

④ Then heat the mixture with stirring for 30 minutes on warm water (50℃). A large number of manganese dioxide has been produced in the process of reaction.

⑤ Suction-filter the mixture, wash the residue with 10mL of 10% sodium carbonate solution.

⑥ Add concentrated sulfuric acid slowly until the solution is strongly acidic and adipic acid has been precipitated. Cool the solution thoroughly collect the crystals of adipic acid by suction filtration and dry it in air. The yield is about 2.2g, the percentage yield is 60.2%, and the melting point is 153℃.

Notes

Method 1

This reaction is itself exothermic. The temperature of the reaction will exceed 45℃ in the initial stage of the reaction. If the temperature does not rise to 45℃ by itself after 5 minutes. carefully heat the mixture to 40℃ in order to make the reaction start.

Method 2

[1] If the water in sodium carbonate solution is less, it will affect the stirring effect and cause potassium permanganate can not sufficiently react.

[2] If the reaction does not start immediately after adding potassium permanganate, carefully heat mixture in a water bath. When the temperature rises to 30℃, remove the warm water bath at once. The exothermic reaction can carry on automatically.

3.10.5 Questions

① Why the temperature of the oxidation reaction must be controlled strictly?

② What is the function of adding sodium carbonate solution to the flask in the method 2?

3.11 Synthesis of Benzoic Acid

3.11.1 Purpose

Master the Principle and method of preparing benzoic acid.

3.11.2 Principle

Benzoic acid and sodium benzoate are important food preservative. Benzoic acid can be used as the pharmaceutical and dye intermediates, also used in the manufacture of plasticizers~initiator in polyester polymerization, spice, etc. In addition, it can be used as the antirust agent of the steel equipment.

Reaction equations

$$C_6H_5CH_3 + 2KMnO_4 \longrightarrow C_6H_5COOK + KOH + 2MnO_2 + H_2O$$

$$C_6H_5COOK \xrightarrow{HCl} C_6H_5COOH + KCl$$

3.11.3 Apparatus and Reagent

(1) Apparatus

Reflux, suction filtration, recrystallization.

(2) Reagent

Toluene, potassium permanganate, sodium bisulfite, Congo-red test paper.

3.11.4 Procedure

① In a 250mL round-bottom flask, place 2.7mL of toluene and 10mL of water. Attach a reflux condenser to the flask and heat the flask over an asbestos gauge until the mixture boils.

② Add in small portions 8.5g of potassium permanganate from the top of the condenser. Finally wash potassium permanganate adhering to the internal wall of the condenser into the flask with 25mL of water.

③ Continue the boiling and shake the flask occasionally until the toluene layer almost disappears and the refluxing liquid appears no longer the droplets of oil (require about 4～5h).

④ Filter the hot reaction mixture under reduced pressure and wash the residue of manganese dioxide with a small mount of hot water (benzoic acid dissolves in hot water, not cold water).

⑤ Combine the filtrate and the washings[1], cool it in an ice-water bath, and then acidify it with concentrated hydrochloric acid until the Congo-red test paper becomes blue and benzoic acid separates out.

⑥ After cooling thoroughly, filter benzoic acid under reduced pressure, wash it with a little cold water, and allow them to dry thoroughly. The yield of the crude product is about 1.7g. The crude product can be purified by recrystallization from water[2].

Pure benzoic acid is a colorless needle-shape crystal, m.p. 122.4℃.

Notes

[1] If the solution is colored purple by residual permanganate, add a small amount of solid sodium bisulfite and filter again under reduced pressure.

[2] If the obtained benzoic acid is not colorless, recrystallize it from appropriate amount of hot water and decolor it with the active charcoal. In 100g of water the solubility of benzoic acid is 0.18g at 4℃, 0.27g at 18℃, and 2.2g at 75℃.

3.11.5 Questions

① Are there any other ways to prepare benzoic acid?

② After the reaction has finished if the filtrate is purple, why should add the sodium bisulfite to the solution?

3.12 Synthesis of Cinnamic Acid

3.12.1 Purpose

① Learn the theory and method of the preparation of cinnamic acid through the Perkin reaction.

② Learn the experimental operation of reflux and steam distillation.

3. 12. 2 Principle

When benzaldehyde is mixed with anhydride and heated in the existence of the correaponding carboxylate, an α, β-unsaturated acid can be prepared through the Perkin reaction.

During the course of this experiment, potassium carbonate can be used to substitute for potassium acetate in the perkin reaction because it will shorten the reaction time and improve the product yield.

$$C_6H_5CHO + (CH_3CO)_2O \xrightarrow[140\sim180℃]{K_2CO_3} C_6H_5CH{=}CHCOOH + CH_3COOH$$

3. 12. 3 Apparatus and reagents

(1) Apparatus

Round-bottom flask, condenser steam distillation apparatus

(2) Reagents

Benzaldehyde, acetic anhydride, anhydrous potassium carbonate, HCl.

3. 12. 4 Procedure

① Place 3mL benzaldehyde, 8mL acetic anhydride, 4. 2g ground anhydrous potassium carbonate and a boiling stone into a 200mL round-bottom flask. Among these materials, benzaldehyde [1] and acetic anhydride[2] are both redistilled before use. Heat the mixture to reflux on the asbestos pad for about 30 minutes[3].

② Cool down the reaction mixture, and add 20mL of water into the flask with shaking or stirring by a glass rod to avoid bulk masses forming. Distill excessive benzaldehyde from the flask by means of steam distillation.

③ Then cool the flask down and add 20mL 10% sodium hydroxide solution as well. Make sure all the cinnamic acid is converted into the sodium salt and dissolves completely.

④ Filter the solution into a 250mL beaker and cool down to room temperature, stir and acidify it with concentrated hydrochloric acid and check it with Congo-red test paper until it becomes blue.

⑤ Filter the solution and wash the solid with an appropriate amount of cold water. Keep the aspirator running until the filter residue becomes dry. Air-dry the crude product. About 3g product can be obtained (65%~75%).

⑥ The crude product can be recrystallized with hot water or the mixture of water and ethanol (the ratio is 3 : 1, V/V). The melting point of pure cinnamic acid is 135. 6℃[4].

Notes

[1] Benzoic acid may be produced by the automatic oxidation of benzaldehyde if it is stored for a long time. Benzaldehyde should be redistilled before use; otherwise the existence of trace benzoic acid not only influences the progress of reaction, but also is difficult to be removed, hence the quality of the product can be ensured. The fraction of 170 ~180℃ is collected and employed as the starting material.

[2] Acetic anhydride may convert into acetic acid when it is exposed to moisture for a long time through hydrolysis and therefore it should be redistilled before as well.

[3] Because carbon dioxide will be released in the process, bubbles will be produced at an early stage of the reaction.

[4] Cinnamic acid has cis and trans isomers, the product prepared usually is transformed with a melting point of 135.6℃.

3.12.5 Questions

What can be get from the reaction of benzaldehyde with propionic anhydride in the presence of anhydrous potassium propionate?

3.13 Synthesis of Ethyl Acetate

3.13.1 Purpose

① Demonstrate direct esterification by the synthesis of ethyl acetate.

② Master the distillation and extraction techniques.

3.13.2 Principle

Ethyl acetate is prepared by direct esterification of the acetic acid with ethanol. The reaction is reversible, and equilibrium is reached through the use of the sulfuric acid catalyst. The yield of the ester can be increased by using an excess of ethanol or by removing one of the products, usually water.

$$CH_3COOH + CH_3CH_2OH \xrightleftharpoons{H_2SO_4} CH_3COOCH_2CH_3 + H_2O$$

3.13.3 Apparatus and Reagents

(1) Apparatus

Three-neck round-bottom flask (125mL), round-bottom flask (50mL), condenser, Erlenmeyer flask, addition funnel (150mL), thermometer (200℃), still heat, distillation adapter.

(2) Reagents

Ethanol (95%), glacial acetic acid, H_2SO_4 (18.4mol/L), saturated aqueous NaCl, saturated aqueous $CaCl_2$, saturated aqueous Na_2CO_3, anhydrous $MgSO_4$, litmus paper.

3.13.4 Procedure

3.13.4.1 Preparation

① Assemble apparatus for the synthesis of ethyl acetate (see Figure 3.3). In a dry

125mL three-neck round-bottom flask, place 6mL of ethanol (95%), Then carefully add 6mL of sulfuric acid (18.4mol/L) dropwise under cooling. Swirl the solution to mix and add two boiling chips. Attach the flask to a reflux assembly using a water cooled condenser. The end of the thermometer and the funnel must immerge into the liquid. Place mixture of 22mL of ethanol (95%) and 18mL of glacial acetic acid into addition funnel (150mL).

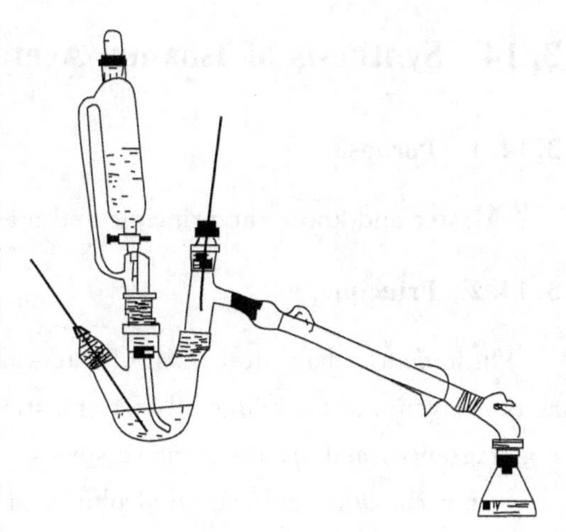

Figure 3.3 Apparatus for the Synthesis of Ethyl Acetate

② Heat the mixture of three-neck round-bottom flask gently with a burner. Then carefully drop the mixture of addition funnel when the temperature rises abut 110℃. To maintain dropping and distillation at the same rate, keep the range 110~125℃. Continue to heat 5~10min until no distillate when the mixture is dropped.

3.13.4.2 The purify

① Slowly add about 10mL saturated aqueous $CaCl_2$ to the distillate [1]. Swirl the mixture gently until carbon dioxide gas is no longer evolved [2]. Transfer the solution to a separatory funnel. Place the separatory funnel on a ring clamp and allow to stand for 5min and discard the lower aqueous layer.

② Draw off. Add 10mL of saturated aqueous NaCl[3] to the ester layer contained in the separatory funnel, stir the mixture gently without shaking: carefully separate the lower aqueous layer and discard it. Wash the upper ester layer with 10mL of saturated aqueous $CaCl_2$, and discard The lower aqueous. Pour the ester layer from the top of the separatory funnel into a dry 50mL of flask, Add about 3g anhydrous $MgSO_4$ to dry the ester about 15min.

③ Rearrange a simple distillation apparatus to distill the ethyl acetate formed. Collect the fraction boiling at approximately 73 ~78℃. Weight the product and calculate percentage yield.

Notes

[1] The crude ester in the distillate contains some acetic acid, which can be removed by neutralization with saturated aqueous Na_2CO_3.

[2] Check to see whether it is basic to litmus. If it is not basic, add additional base until the solution is basic.

[3] Wash with saturated aqueous NaCl to aid in lay separation.

3.13.5 Questions

① What is the purpose of the use of sulfuric acid in this experiment?

② What methods can be used to the equilibrium in the esterification in order to favor products?

③ Which impurities may be contained in the crude ester, and how can they be removed?

3.14 Synthesis of Isoamyl Acetate (banana oil)

3.14.1 Purpose

Master and know the principle and method of preparing isoamyl acetate.

3.14.2 Principle

In human's daily life, most of esters have the widespread use. Some esters may be used as the solvents of the edible oil, fat, plastics and paint. Many esters have the pleasant-smelling fragrance, and are inexpensive spices.

An carboxylic acid and an alcohol or a phenol react in the presence of an inorganic or a organic acid catalyst to produce ester and water, this process is called esterification reaction. The commonly used catalysts are concentrated sulfuric acid, dry hydrogen chloride, organic acid or cation exchange resin. The esterification reaction can also achieve in the absence of a acid catalyst if the ability of the carboxylic acid participating reaction is stronger, such as formic acid, oxalic acid.

The esterification reaction is reversible. If equimolar amounts of acid and alcohol are used and the equilibrium has been established, the yield is only 67%. To increase the mount of ester formed, the use of an excess of the alcohol or organic acid will make the equilibrium shift to the right. The choice of reactant to be used in excess will depend upon factors such as availability, cost, and ease of removal of excess reactant from the product.

In addition, driving an esterification to completion by removal of the water or ester formed in the reaction is a common practice. Water formed in the reaction can be removed by azeotropic distillation using, for example, benzene, toluene or chloroform, etc. As the solvent, which forms an azeotrope with water. Ester may be also prepared by the alcoholysis of acid chlorides or acid anhydrides.

Reaction:

$$CH_3COOH + HOCH_2CH_2\underset{\displaystyle CH_3}{\underset{|}{C}}HCH_3 \overset{H^+}{\rightleftharpoons} CH_3COOCH_2CH_2\underset{\displaystyle CH_3}{\underset{|}{C}}HCH_3 + H_2O$$

3.14.3 Apparatus and Reagents

(1) Apparatus

Reflux, extraction/separation, distillation apparatus; Abbe refractometer.

(2) Reagents

Isoamyl alcohol, glacial acetic acid, concentrated sulfuric acid, 5% aqueous sodium carbonate, sodium chloride, anhydrous magnesium sulfate.

3.14.4 Procedure

① Pour 5mL of isoamyl alcohol and 7mL of glacial acetic acid into a dry 25mL round-bottom flask. Slowly add 1.0mL of concentrated sulfuric acid[1] to the contents of the flask with swirling. Add two boiling stones to the mixture. Fit a reflux condenser to the flask, and heat the mixture under reflux for 1.0h.

② Then, remove the heating source and allow the mixture to cool to room temperature. Pour the cooled mixture into a separatory funnel.

③ Rinse the reaction flask with 10mL of cold water and pour the rinsing into the separatory funnel. Stopper the funnel and shake it several times.

④ Separate the lower aqueous layer, and wash the organic layer with 5mL of 5% aqueous sodium carbonate[2], then with 5mL of saturated aqueous sodium chloride until the organic layer is neutral[3].

⑤ Separate the layers and discard the aqueous layer. Transfer the crude ester to a clean, dry 25mL flask and dry it over anhydrous magnesium sulfate.

⑥ Filter the dried product into a drying distilling flask, distill and collect the fraction boiling at 138～142℃ in a dry flask which has been immersed in an ice bath to ensure condensation and to reduce odors.

⑦ Weigh the product, calculate the percentage yield and determine the refractive index.

Pure isoamyl acetate is a colorless, transparent liquid, b. p. 138～142℃, d_4^{20} 0.876, n_D^{20} 1.4000.

Notes

[1] The reaction liquid is exothermal while adding concentrated sulfuric acid to the flask. Carefully shake the flask to rapidly diffuse the caloric.

[2] When the crude product is washed by the sodium carbonate solution, carbon dioxide gas will evolve. Do not shake the funnel acutely, and release the inner pressure of the funnel.

[3] Wash the organic layer with saturated aqueous sodium chloride until it is neutral. Other wise the product is easily decomposed during distilling.

3.14.5 Questions

① One method for favoring the formation of the ester is to add an excess of acetic acid. Suggest another method involving the right-hand side of the equation, which will favor the formation of the ester.

② Why is excess acetic acid easier to remove from the products than excess isoamyl alcohol?

3.15 Synthesis of Acetylsalicylic Acid (aspirin)

3.15.1 Purpose

① Learn and master the recrystallization procedure.

② Mater the method of determining acetylsalicylic acid.

3.15.2 Principle

Acetylsalicylic acid (aspirin) can be usually synthesized by acylation of salicylic acid. In this experiment, acetic anhydride is used as an acylating agent and concentrated sulfuric acid is used as a catalyst. The action of H_2SO_4 is to break the intramolecular hydrogen bonding between carboxyl (—COOH) and hydroxy (—OH) in salicylic acid so as to accelerate the reaction.

main reaction

$$\text{C}_6\text{H}_4(\text{OH})\text{COOH} + (CH_3CO)_2O \xrightleftharpoons{H^+} \text{C}_6\text{H}_4(\text{OCOCH}_3)\text{COOH} + CH_3COOH$$

The crude product may contain a small amount unreacted salicylic acid which can be removed by recrystallization technique.

The purity of the sample will be tested with the solution of $FeCl_3$ and compared with a commercial aspirin and salicylic acid.

The content of acetylsalicylic acid will be used to determinate by UV-751 Spectrophotometer

3.15.3 Apparatus and Reagents

(1) Apparatus

Round-bottom flask (150mL), water bath pot, thermometer, condenser, beaker (100、50mL), Buchner funnel, hot funnel.

(2) Reagents

Salicylic acid, acetic anhydride, H_2SO_4 (conc.), ethanol (95%), $FeCl_3$ (0.006mol/L).

3.15.4 Procedure

3.15.4.1 Preparation

① Assemble the apparatus of the synthesis of acetylsalicylic acid, see Figure 2.14.

② Place 5.0g of Salicylic acid crystals in a 150mL round-bottom flask. Add 10mL of acetic anhydride followed by 5 drops of concentrated H_2SO_4, and stir carefully with a glass rod.

③ Heat the flask on the hot water bath (70～80℃) for 30 minutes, and shake the flask sometimes during the reaction.

④ Cool the mixture for a while after the reaction has completed, then pour the mixture into a 100mL beaker containing 30mL cold water while stirring. Cool the mixture in an ice-cold water bath until the crystallization has completed.

⑤ Collect the product by suction on a Buchner funnel and wash the crystals with cold water twice, then the crude product is obtained.

3.15.4.2 Purification

① Transfer the crude product to a 100mL beaker and add 6mL of ethanol (95%), heat gently on the hot water bath (60～70℃) until solid completely dissolve, then hot filter.

② Add 15mL of distillation water to filtrate while stirring.

③ Cool the beaker in a ice-cold bath until the crystallization has completed.

④ Collect the product by suction on the Buchner funnel, wash the solid with cold ethanol (95%) twice.

⑤ Transfer the solid to a glass watch, drying the solid.

⑥ Calculate the percentage yield.

3.15.4.3 Identify the Purity

Dissolve a few crystals of salicylic acid or the crude product in two test tubes each of them containing 5 drops of ethanol, add 2 drops of ferric chloride solution to each tube and compare the color.

3.15.5 Questions

① When we prepare acetylsalicylic acid, the apparatuses used must be anhydrous, why?

② Why do we use the coned H_2SO_4 in the acetylation reaction?

③ What side reaction will occur in the experiment?

3.16 Synthesis of Methyl Salicylate (oil of wintergreen)

3.16.1 Purpose

① Learn the principle of preparing methyl salicylate (oil of wintergreen),

② Master the method of preparing methyl salicylate.

3.16.2 Principle

Methyl salicylate was first isolated in 1843 by extraction from the wintergreen plant (Gaultheria). Therefore it is called oil of wintergreen. It is a natural ester, exists in ylang-ylang oil, sesame oil and clove oil. It has a pleasant odor from the leaves of the wintergreen plant, and it is usually used as a flavoring. It is frequently used for food, toothpaste and cosmetics and so on. It also has analgesic and antipyretic character. Methyl salicylate may be prepared by esterifying salicylic acid with methyl alcohol, a reaction in which the catalyst concentrated sulfuric acid.

Reaction equation

$$o\text{-}HOC_6H_4COOH + CH_3OH \xrightarrow{H_2SO_4} o\text{-}HOC_6H_4COOCH_3 + H_2O$$

3.16.3 Apparatus and Reagents

(1) Apparatus

Reflux, distillation, extraction/separation, vacuum distillation apparatus; instrument Abbe refractometer.

(2) Reagents

Salicylic acid, methanol, concentrated sulfuric acid, sodium bicarbonate.

3.16.4 Procedure

① Set up a reflux apparatus, using a dry 30mL round-bottom flask[1]. Add 3.45g of salicylic acid and 15mL of methanol to the flask. Swirl the flask to dissolve the salicylic acid. Slowly add 1mL of concentrated sulfuric acid to the mixture while cooling and swirling to thoroughly mix the reactants[2]. Add two boiling stones to the flask, and assemble the apparatus.

② Heat the mixture under reflux for 1.5h.

③ Remove the condenser and reassemble the apparatus for distillation. Distill off the excess methanol in water bath.

④ Cool the residual mixture in the reaction flask, and then add 10mL of water. After swirling, transfer the mixture to a separatory funnel. Separate the organic layer.

⑤ Extract the aqueous layer with 10mL of diethyl ether. Combine the extract with the organic layer.

⑥ Wash the ether solution with 10mL of water, then with 10mL of 10% sodium bicarbonate solution, and finally with water until the organic layer shows neutral.

⑦ Dry the organic layer containing the product with anhydrous magnesium sulfate.

⑧ Filter off the drying agent and distill off diethyl ether in water bath[3]. Distill the crude product by vacuum distillation. Collect the fraction boiling at 115～117℃/2.7kPa (20 mmHg) in a weighed bottle. Weigh the product, calculate the percentage yield, and determine the refractive index.

Pure methyl salicylate is a colorless and transparent liquid, m.p. −8～−7℃, b.p. 222℃, d_4^{25} 1.1787, n_D^{20} 1.5360.

Notes

[1] All apparatus must be dried thoroughly before use. Otherwise the yield of the esterification will be reduced.

[2] If the flask does not be shaken in time while dropping concentrated sulfuric acid, sometimes the partial reactants may be carbonized.

[3] Distill off methanol thoroughly, or the yield will be reduced because of increasing the solubility of the product in water.

3.16.5 Questions

① Salicylic acid is esterified with methanol group to yield methyl salicylate (oil of wintergreen). The esterification is an acid-catalyzed equilibrium reaction, what measures has been taken in this experiment in order that the equilibrium moves toward the right to favor the formation of the ester in high yield? What methods can be also adopted?

② If the methanol was not distilled off, rather than the crude product was washed directly with water after the esterification reaction had been finished, what influence would it have on the experimental result?

③ Why was 10% $NaHCO_3$ used to wash the crude product in the final treatment? What would have happened if the aqueous sodium hydroxide solution had been used?

④ What is the function of the sulfuric acid in this reaction? Write a mechanism for the acid-catalyzed esterification of salicylic acid with methanol.

3.17 Synthesis of Ethyl Acetoacetate

3.17.1 Purpose

① Learn the principle and method of preparing ethyl acetoacetate by Claisen condensation of ethyl acetate.

② Master the technique of vacuum distillation.

3.17.2 Principle

Ethyl Acetoacetate is usually prepared by the Claisen condensation of ethyl acetate.

$$2CH_3COOC_2H_5 \xrightleftharpoons{C_2H_5ONa} CH_3\overset{O}{\overset{\|}{C}}CH_2\overset{O}{\overset{\|}{C}}OC_2H_5 + C_2H_5OH$$

3.17.3 Apparatus and reagents

(1) Apparatus

Reflux, vacuum distillation apparatus.

(2) Reagents

Na, ethyl acetate, acetic acid, saturated aqueous NaCl, Na_2CO_3, anhydrous magnesium sulfate.

3.17.4 Procedure

① In a dry 50mL round-bottom flask[1], place 9.8mL of anhydrous ethyl acetate, then quickly add 1g of sodium hydroxide-free sodium metal as thin flakes[2,3].

② Place the flask in a reflux apparatus with a warm-water bath as the heat source. At the top of the condenser fit a drying tube containing anhydrous calcium chloride.

③ Heat the reaction mixture under reflux gently until the sodium metal disappears (about 2 hours). The solution will be clear red.

④ Turn off the heating source and allow the mixture to cool[4].

⑤ Add 50% acetic acid[4] until the solution is just slightly acid to blue litmus paper (pH =5~6) with stirring slowly: a solid may appear, but it dissolves in acid conditions.

⑥ Transfer the reaction mixture to a separatory funnel, add an equal volume of saturated aqueous sodium chloride, shake it thoroughly and let it stand for a few minutes.

⑦ Separate the ester layer, extract the aqueous layer with 5mL of ethyl acetate and mix the extract with the ester layer.

⑧ Wash the organic layer with 5% Na_2CO_3 until the solution is neutral to pH paper.

⑨ Save the organic layer in a small flask, dry the crude product using small amount of anhydrous magnesium sulfate.

⑩ Decant the clean liquid into a dry distilling flask. Remove the ethyl acetate from the ethyl acetoacetate by distilling on a water-bath.

⑪ Set up an apparatus for reduced pressure distillation. Vacuum-distill the product at about 931Pa (7mmHg), at which pressure the product is collected between 54 ～55℃.

⑫ Weight the product and calculate the percentage yield.

Pure ethyl acetoacetate is a colorless liquid, b. p. 180. 4℃ (decomposition), d_D^{20} 1. 0282, n_D^{20} 1. 4194.

Notes

[1] All the glassware used in this experiment must be thoroughly dry and the reagents must be absolutely anhydrous.

[2] It is necessary to remove any oxide at the surface of the sodium and cut the sodium into small pieces for speeding up the condensation.

[3] The reaction starts immediately. If it gets too violent, immerse the flask in the cold water bath to stem the reaction. If the reaction does not start, heat the reaction mixture to promote the reaction, and then, remove the heat source.

[4] Be careful when adding 50% acetic acid, because it reacts violently with unreacted sodium.

3. 17. 5 Questions

① Why must all the glassware and reagents used in this experiment be dry and absolutely anhydrous?

② In the work up procedure ethyl acetoacetate, what is the purpose of adding 50% acetic acid and saturated NaCl?

3. 18 Synthesis of Acetanilide

3. 18. 1 Purpose

① To acquaint the method and principle of acetanilide's synthesizing via acetylation reaction.

② Master the operation skill of organic compound's reflux and recrystallization.

3. 18. 2 Principle

The acylation reaction is carried out to synthesize acetanilide with aniline and acetic acid as the starting materials in the presence of zinc powder.

$$C_6H_5NH_2 + CH_3COOH \rightleftharpoons C_6H_5-NH-C(=O)-CH_3 + H_2O$$

Pure acetanilide, a white plate crystal (m. p. 114℃), can hardly dissolve in cold water, and acetanilide solid can be dissolved slightly by hot water[1]. So hot water can be used as the crystallization solvent of acetanilide. In this experiment, excess acetic acid is used as the acetylation reagent.

3.18.3 Apparatus and Reagents

(1) Apparatus

Round-bottom flask (150mL), condenser; Vigreux fractionating column, safe bottom, thermometer (150℃), beaker (250mL), a pair of scissors, decompression-filtration flask, Buchner funnel, vacuum pump; water-bath, electricfurnace, filter paper.

(2) Reagents

Aniline (CP), acetic acid (CP), zinc dust, activated charcoal.

3.18.4 Procedure

3.18.4.1 Preparation

① Assemble the apparatus of the synthesis of acetanilide (see Figure 1.7).

② The reaction is performed in a 150mL of round-bottom flask, 5mL of aniline distilled recently, 7.5mL of acetic acid and zinc powder (about 0.1g)[2] are added in the reaction flask respectively, then the reactor is heated about 1 hours during boiling and maintaining temperature 105℃. The reaction is stopped until it appears white smog.

③ The hot reaction mixture is poured into a beaker which contains 100mL of cold distillation water, of course, the mixed solution should be stirred constantly. After cooling, the crude product will crystallize out. When there is no more crystalline out of the solution, the crude product is filtered in a decompression filtration apparatus with a vacuum pump to decompress, and then washed by 5～10mL of glacial distillation water to clear the rudimental acid solution away from the product[3].

3.18.4.2 Purification

① The crude acetanilide is dissolved by 100mL of hot distillation water, followed by heating to dissolve the indissoluble oil. After the solution is boiling for a while, if there is still indissoluble oil, it needs more hot distillation water to be added in the solution until the oil dissolves completely[4].

② The next step is to stop beating and place to cool it naturally far several minutes. Then, about 1g of activated charcoal is added in the solution[5], which followed by stirring with a glass rod and decolorizing for 5 minutes. After that, the hot decoloration solution is filtered immediately by the hot funnel [see Figure 1.11 Heat Filtration (a) or (b)].

③ The filtrate is cooling and acetanilide will crystallize out from the filtrate. When there is not crystal producing any more, the product is collected with the method of decompression filtration. Then it needs washing repeatedly twice with a little cold distillation water and pressing by a glass rod and filtering with the decompression-filter to eliminate water as. After that, the product, acetanilide, is collected and placed in a watch glass to place naturally temperature is not beyond 60℃. The final product is weight to calculate the yield.

The melting-point of pure acetanilide solid is 114.3℃.

Notes

[1] The solubility of acetanilide in 1L of water: 0.0458mol under 25℃, 0.0683mol under 60℃, 0.2846mol under 80℃, 0.4588mol under 100℃.

[2] Zinc powder can prevent aniline against being oxidized in the reaction. Excess zinc powder shouldn't be added in reaction solution, otherwise there would be a little indissoluble product $Zn(OH)_2$ to yield in the next steps.

[3] After the reaction products are cooled down, the solid product will crystallize out and cling to the wall of the reactor, which will be a trouble to the next step. To avoid appearing this phenomenon, the hot reaction should be poured into cold distillation water and the mixture should be stirred to separate the excess acetic acid and the remained aniline from the solid product.

[4] The oil is not impurity, but acetanilide. Because when the solution temperature is beyond 83℃, there will be a little acetanilide solid that not dissolve in water, melts and exists on oil formation in this condition.

[5] Adding activated charcoal in boiling solution can result in the violent-boiling of the container. So, therefore activated charcoal is added, tie hot solution should be placed naturally for several minutes.

3.18.5 Questions

① what kind of compounds is often used as acetylating reagents in organic reaction? Which one is more economical?

② which method can be used to purify solid organic compounds? How to identify the purification of compound with a simple method?

3.19 Synthesis of *p*-toluidine

3.19.1 Purpose

Master the Principle and method of preparing *p*- toluidine.

3.19.2 Principle

Reaction:

$$CH_3C_6H_4NO_2 \xrightarrow[H^+]{Fe} CH_3C_6H_4NH_2$$

3.19.3 Apparatus and Reagents

(1) Apparatus

Mechanical stirrer; apparatus for: reaction with reflux, suction filtration, distillation, extraction/separation.

(2) Reagents

Reductive ferrum powder, *p*-nitrotoluene, benzene, sodium bicarbonate, hydrochloric acid, ammonium chloride, zinc powder.

3.19.4 Procedure

① Place 7.00g of reductive ferrum powder, 0.90g of ammonium chloride and 20mL of water[1] in a 50mL three-neck flask equipped with a reflux condenser and a mechanical stirrer. Heat the mixture while stirring and boil it gently over a small flame for 15 minutes.

② Cool the flask slightly, and then add 4.50g *p*-nitrotoluene to the flask. Continue the reflux while stirring for 1 h.

③ After the reaction is finished, cool the mixture to room temperature and neutralize it with 5% sodium bicarbonate solution[2].

④ Add appropriate amount of benzene to the mixture with stirring. Remove the residue of ferrum powder[3] by suction filtration and wash the residue with a little benzene.

⑤ Pour the filtrate into a separatory funnel, separate the benzene layer, extract the aqueous phase three times with benzene and then combine the extracts.

⑥ Extract again the extracts of benzene three times with 5% hydrochloric acid and combine the extracts of hydrochloric acid.

⑦ Add 20% sodium hydroxide solution to the extract of hydrochloric acid with stirring. The crude product separates.

⑧ Collect the obtained crude product by suction filtration, wash it with a little water and re-extract the aqueous phase with a little benzene. Combine the extract with the crude product.

⑨ Distill off the benzene from a water bath. Add a small amount of zinc powder to the residue and distill the residue over an asbestos gauge[4]. Collect the fraction boiling at 198～201℃. The yield is about 2.50g, m p. 44～45℃.

Pure *p*-toluidine is a colorless sheet-shape crystal, m. p. 44～45℃, b. p. 200.3℃. *p*-Toluidine becomes black easily in air and light because of the oxidation.

Notes

[1] In this experiment, ferrum-hydrochloric acid is used as the reducer, in which the hydrochloric acid is obtained by the hydrolysis of ammonium chloride.

[2] Control the pH value between 7 and 8 while adding sodium bicarbonate to the reaction mixture to avoid the formation of gelatinous ferric hydroxide under the condition of stronger alkaline, which will make the separation difficult.

[3] The ferrum residue is the active ferrum mud, containing 44.7% divalent ferrum (calculated by FeO), the black particles. It will liberate heat vigorously in air, thus it should be poured into the waste tank containing water in time.

[4] *p*-Toluidine can be also purified by the recrystallization from ethanol-water except distillation.

3.19.5 Questions

① Before the reduction reaction, why is the ferrum powder pretreated?

② During the treatment, why are the sodium bicarbonate solution and benzene added first, and then extract the layer of benzene with 5% hydrochloric acid?

3.20 Synthesis of Anthranilic Acid (2-aminobenzoic acid)

3.20.1 Purpose

Master the principle and method of preparing anthranilic acid.

3.20.2 Principle

Reactions

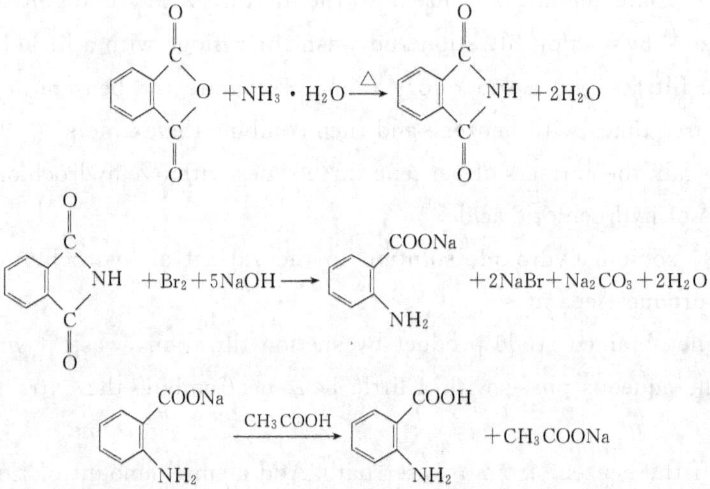

$$\text{(phthalic anhydride)} + NH_3 \cdot H_2O \xrightarrow{\triangle} \text{(phthalimide, NH)} + 2H_2O$$

$$\text{(phthalimide, NH)} + Br_2 + 5NaOH \longrightarrow \text{(COONa, NH}_2\text{)} + 2NaBr + Na_2CO_3 + 2H_2O$$

$$\text{(COONa, NH}_2\text{)} \xrightarrow{CH_3COOH} \text{(COOH, NH}_2\text{)} + CH_3COONa$$

Reaction mechanism

$$\text{(phthalimide, NH)} + Br_2 + 5NaOH \longrightarrow \text{(COONa, NH}_2\text{)} + 2NaBr + Na_2CO_3 + 2H_2O$$

$$\text{(phthalimide)} \xrightarrow{NaOH} \text{(COONa, CONH}_2\text{)} \xrightarrow{NaOBr/NaOH} \left[\text{(COONa, CONHBr)}\right] \xrightarrow{-HBr} \left[\text{(COONa, CON:)}\right] \longrightarrow \left[\text{(COONa, N=C=O)}\right] \xrightarrow{H_2O} \text{(COONa, NH}_2\text{)}$$

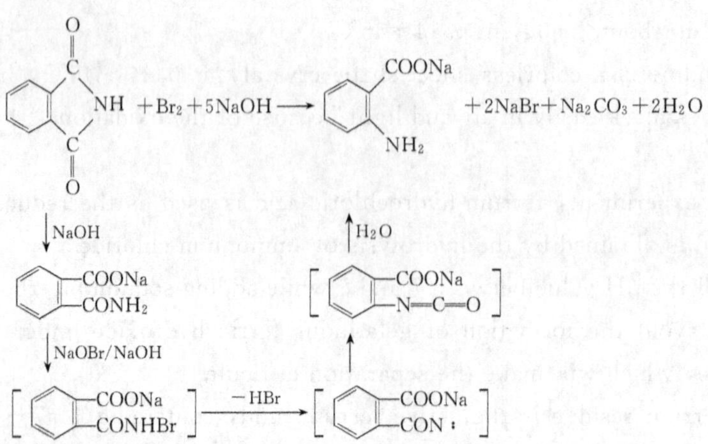

3.20.3 Apparatus and Reagents

(1) Apparatus

Apparatus for: reflux, suction filtration, recrystallization.

(2) Reagents

Phthalic anhydride, ammonia water, bromine, sodium hydroxide, hydrochloric acid, saturated sodium bisulfite, glacial acetic acid, litmus paper.

3.20.4 Procedure

(1) Preparation of Phthalimide

Place 10g of phthalic anhydride and 10mL of concentrated ammonia water in a 100mL two-neck flask. Fit the flask with an air condenser and a 360℃ thermometer. First heat the flask over an asbestos gauge. then heat the flask directly with a small flame while shaking occasionally, until the reaction temperature rises gradually to 300℃. Push the solid formed by sublimation in the condenser into a porcelain dish. Pout the hot reaction product into a porcelain dish. After cooling, place the concretionary solid in a mortar and grinds it to powder to be used. The yield is about 8g, m. p. 232～234℃.

(2) Preparation of Anthranilic Acid

In a 125mL flask dissolve 7.5g of sodium hydroxide in 30mL of water and cool the solution to −5～0℃ in an ice-salt bath. Add 2.1mL of bromine[1] in one portion to the basic solution and shake the flask vigorously until all of the bromine has reacted. At the moment the reaction temperature will rise slightly. Cool the sodium hypobromite solution below 0℃ to be used.

In another small flask dissolve 5.5g of sodium hydroxide in 20mL of water to prepare another basic solution.

Add the pasty mass prepared by 6g of the finely powdered phthalimide and a small amount of water to the cold sodium hypobromite solution while shaking the flask vigorously and maintaining the temperature of the reaction mixture at about 0℃. Remove the ice-salt bath and continue shaking the flask until the reactant changes into the yellow clear solution. Rapidly add the prepared sodium hydroxide solution to the flask. The temperature of the mixture rises spontaneously. Heat the mixture to 80℃ for about 2 minutes. Add 2mL of saturated sodium bisulfite solution to reduce the residual sodium hypobromite. Cool the solution and filter under reduced pressure. Pour the filtrate to a 250mL beaker, cool the solution in an ice-water bath and add concentrated hydrochloric acid dropwise to the solution with stirring until the solution is just neutral (pH=7, ca. 15mL should be necessary, test by litmus paper)[2]. Precipitate the 2-aminobenzoic acid by slow addition of glacial acetic acid (ca. 5～7mL)[3], filter the precipitate with suction, wash it with a little cold water and dry it in air. The yield is about 4g. Recrystallize the grey crude product from water. The white sheet-shape crystals are obtained.

Pure 2-aminobenzoic acid is a white sheet-shape crystal, m. p. 145℃.

Notes

[1] Bromine is highly corrosive and irritant. Always wear goggles and rubber gloves and measure out in the hood. Do not inhale the bromine vapor.

[2] Anthranilic acid can dissolve both in the alkali and in the acid, therefore the exces-

sive hydrochloric acid will make the product dissolved. If too much acid is added, neutralize the mixture by sodium hydroxide solution.

[3] The isoelectric point of anthranilic acid is about 3～4. In order to isolate anthranilic acid completely, add the right amount acetic acid to the solution.

3.20.5 Questions

① Which applications does anthranilic avid have in the synthesis and the analysis?

② If the used bromine and the sodium hydroxide are insufficient or more excessive, what is the influence on this reaction respectively?

③ When the basic solution of anthranilic acid is just neutral with adding hydrochloric acid to the solution, why should the right amount of acetic acid be added rather than the hydrochloric acid to isolate the product completely?

3.21 Synthesis of Phenoxyacetic Acid

3.21.1 Purpose

Master the principle and method of preparing phenoxyacetic acid.

3.21.2 Principle

Phenoxyacetic acid is a white flake crystal or needle-shape crystal. It can be used for the synthesis of dye, medicine and pesticide. It can be directly used as a plant growth regulator which is harmless to humans and animals. Phenoxyacetic acid has been widely applied. It can be prepared by the Williamson reaction of phenol with chloroacetic acid in aqueous alkali.

Overall reaction equation

$$C_6H_5OH \xrightarrow[OH^-]{ClCH_2COOH} C_6H_5OCH_2COOH$$

Stepwise reaction equation

$$C_6H_5OH + NaOH \longrightarrow C_6H_5ONa + H_2O$$

$$2ClCH_2COOH + Na_2CO_3 \longrightarrow 2ClCH_2COONa + H_2O + CO_2$$

$$C_6H_5ONa + ClCH_2COONa \longrightarrow C_6H_5OCH_2COONa + NaCl$$

$$C_6H_5OCH_2COONa + HCl \longrightarrow C_6H_5OCH_2COOH + NaCl$$

3.21.3 Apparatus and Reagents

(1) Apparatus

Motor stirrer, reflux with stirring apparatus, extraction/separation, suction filtration.

(2) Reagents

Chloroacetic acid, phenol, sodium hydroxide, hydrochloric acid, diethyl ether, 15% sodium chloride, sodium carbonate.

3.21.4 Procedure

(1) Preparation of Sodium Chloroacetate Solution

Add 3.1g of chloroacetic acid[1] and 10mL of 15% sodium chloride solution to a 100mL beaker by turn, then add about 2g of sodium carbonate to the beaker while stirring, at such a rate that the temperature of the reaction solution does not surpass 40℃[2]. After all of the sodium carbonate has been added, the pH of the solution is 7~8. If not, adjust that until 7~8 with saturated sodium carbonate solution.

(2) Preparation of Sodium Phenate Solution

Fit a 100mL three-neck flask with a motor stirrer[3], reflux condenser and thermometer. Dissolve 2.8g of phenol and 1.3g of sodium hydroxide in 7.5mL of water in the flask with stirring. Then cool the solution to the room temperature.

(3) Preparation of Phenoxyacetic Acid

Pour the prepared sodium chloroacetate solution into the flask containing sodium phenate solution. Heat the flask gently over an asbestos gauge with a small flame while stirring for 2h and keep the temperature of the reaction in the rang of 100~110℃[4].

After the reaction has finished, pour the hot reaction mixture into a 250mL beaker, add 30mL of water to the beaker, stir uniformly, adjust the pH of the solution between 1~2with concentrated hydrochloric acid. In this process, the crude phenoxyacetic acid separates in white crystals. After cooling, collect the white crystals with suction filtration and wash them with 5mL of cold water. Place the crude product into a 250mL beaker, add 30mL of water and sodium carbonate while stirring until the crude product is dissolved. Transfer the solution to a separatory funnel, and then extract the aqueous solution with 10mL diethyl ether[5] to remove the unreacted phenol. Separate the aqueous layer and acidify it with 20% hydrochloric acid until the pH of the solution is 1~2. Allow the contents to cool, collect the white crystals with suction filtration and wash them with a little cold water twice. After drying, the pure phenoxyacetic acid is obtained. Weigh it, determine the melting point of it and calculate the yield.

Pure phenoxyacetic acid is a white and needle-shape crystal, m.p. 98~99℃.

Notes

[1] Monochloroacetic acid and phenol is corrosive, avoid our skin being touched.

[2] Preparing the solution of sodium chloroacetate, the use of the aqueous sodium chloride solution is favourable to inhibit the hydrolysis of sodium chloroacetate. When the temperature of the solution is over 40℃ during neutralization, the sodium chloroacetate hydrolyzes easily.

[3] Install and fix the motor stirrer carefully to avoid damaging the glassware.

[4] When the reaction of preparing phenoxyacetic acid has just started, the pH value of

the reaction mixture is 12. The pH value will fall gradually along with the reaction until it is 7～8. At this time, the reaction finishes.

[5] Diethyl ether is used to remove the unreacted phenol.

3.21.5 Questions

① When the ether, phenoxyacetic acid, is prepared by sodium phenate and Monochloroacetic acid, why should the Monochloroacetic acid be made into sodium chloroacetate? Whether the ether can be prepared by phenol and Monochloroacetic acid directly?

② Why is the aqueous sodium chloride solution added when Monochloroacetic acid is neutralized by sodium carbonate?

③ Why does the pH value of the solution change in the course of preparing phenoxyacetic acid? Why is the reaction complete when the pH value of the reaction solution is 7～8?

3.22 Synthesis of Methyl Orange

3.22.1 Purpose

① Learn the principle and method of preparing methyl orange from sulfanilic acid and *N,N*-dimethylaniline by the diazo coupling reaction.

② Master the technique of recrystallization.

3.22.2 Principle

Methyl orange is commonly used as an acid-base indicator. In this experiment it will be prepared by the diazo coupling reaction from diazotized sulfanilic acid and *N,N*-dimethylaniline.

$$H_2N-C_6H_4-SO_3H \xrightarrow{NaOH} H_2N-C_6H_4-SO_3Na \xrightarrow[0\sim5^\circ C]{NaNO_2+HCl} HO_3S-C_6H_4-N_2Cl \xrightarrow[HAc]{C_6H_5-N(CH_3)_2}$$

$$HO_3S-C_6H_4-N{=}N-C_6H_4-N(CH_3)_2 \xrightarrow{NaOH} NaO_3S-C_6H_4-N{=}N-C_6H_4-N(CH_3)_2$$

Methyl Orange

3.22.3 Apparatus and reagents

(1) Apparatus

Beaker, hot funnel, Buchner funnel.

(2) Reagents

Sulfanilic acid, *N,N*-dimethylaniline, sodium nitrite, NaOH, HCl, glacial acid, NaCl.

3.22.4 Procedure

3.22.4.1 Diazotization

① Place 2.1g of sulfanilic acid into a 50mL beaker, add 10mL of 5% aqueous sodium hydroxide solution[1].

② Stir the mixture with a glass rod and warm it in a hot water bath until it dissolves.

③ Allow the solution to cool to room temperature, add 0. 8g of sodium nitrite and stir until it dissolves.

④ Pour slowly this solution batchwise, with stirring[2], into a 250mL beaker containing 13mL of water and 2. 5mL of concentrated hydrochloric acid. During the reaction the temperature should be controlled below 5℃[3].

⑤ After adding, test the reaction solution with a starch-potassium iodide test paper[4]. Keep the mixture in an ice-salt bath for about 15min to complete the reaction.

3. 22. 4. 2 Coupling Reaction

① In a test tube containing 1. 3mL of *N*, *N*-dimethylaniline and 1mL of glacial acetic acid, mix the solution thoroughly by swirling the test tube.

② Slowly, with stirring, add this solution to the cooled diazotized sulfanilic acid in the 250mL beaker. Stir the mixture vigorously with a stirring rod.

③ Keep the mixture in an ice bath for about 15min to make sure the coupling reaction is completed.

④ Add 15mL of a 10% aqueous sodium hydroxide solution[5]. Check with litmus or pH paper to guarantee the solution is basic. If not, add extra base. The color of solution turns to orange. The crude methyl orange is precipitated with fine-granular shape.

⑤ Heat the mixture with a boiling water bath for 5min to dissolve most of the newly formed methyl orange.

⑥ When all (or most) of methyl orange is dissolved, add 5g of sodium chloride, cool the mixture to room temperature and put it in an ice bath. The methyl orange will restart to be crystallized and isolated.

⑦ When the precipitation appears completely, collect the product by vacuum filtration. using a Buchner funnel[6].

⑧ Rinse the beaker with two cold portions of a saturated sodium chloride solution (5mL each) and wash the filter cake with these rinse solutions.

⑨ Purify the product by recrystallizing the crude methyl orange from boiling water.

⑩ Allow the product to dry, weigh it, and calculate the percentage yield (about 75%). Pure methyl orange is an orange slice-shaped crystal, with no definite melting point, pH 3. 1 (red)~pH 4. 4 (orange).

Notes

[1] Sulfanilic acid, insoluble in acid solution and soluble in basic solution, is an amphoteric compound. Its acidity is slightly stronger than its basicity. It is nevertheless necessary to carry out diazotization reaction in an acid (HNO_2) solution. To resolve this problem, sulfanilic acid is first dissolved in basic solution, not in acidic solution.

[2] Continuous stirring is important to this diazotization. In a short time the diazonium salt of sulfanilic acid should separate as a finely divided white precipitate.

[3] Temperature is important in this reaction. If the temperature of the reaction is above 5℃, the diazonium salt will be hydrolyzed to a phenol, thus reducing the yield of the desired product.

[4] At this moment the solution is weak alkaline. If neutral, the small amounts of alkaline liquid should be added until the solution just appears alkaline. If strongly alkaline, the resin-like polymer, not desired product, will be formed easily.

[5] The color of wet methyl orange can deepen quickly when exposed in the natural light. So the ordinary crude methyl orange is a red-purplish (mauve) crystal. It will dry quickly if washed first with alcohol and then ether.

[6] The product is salt. Since salts do not generally have well-defined melting points, the determination of melting point should not be attempted.

3.22.5 Questions

① In this experiment, why should the temperature of procedure be controlled below 5℃?

② Why is the dimethylaniline coupling with diazonium salt formed a para-position product?

③ What would the result be if cuprous chloride were added to the diazonium salt prepared in this reaction?

3.23 Synthesis of Benzocaine

3.23.1 Purpose

① Master the method and principle of preparing benzocaine.

② To know experimental technique of the multistep synthesis.

3.23.2 Principle

Benzocaine is common name of ethyl *p*-aminobenzoate. It is used as local anesthetics or painkiller. There are two steps for the synthesis of benzocaine. First, it is reduction of *p*-nitrobenzoic acid, second, it is esterification of *p*-aminobenzoic acid.

Reduction:

Sn is reductant, in the presence of HCl. $SnCl_4$ is removed by add $NH_3 \cdot H_2O$ to form the precipitate after reduction. Then adjusted pH with $NH_3 \cdot H_2O$ and ice-acetic acid.

$$p\text{-}O_2N\text{-}C_6H_4\text{-}COOH \xrightarrow[HCl]{Sn} p\text{-}(NH_2 \cdot HCl)\text{-}C_6H_4\text{-}COOH + SnCl_4$$

$$SnCl_4 + 4NH_3 \cdot H_2O \longrightarrow Sn(OH)_4 \downarrow + 4NH_4Cl$$

$$p\text{-}(NH_2 \cdot HCl)\text{-}C_6H_4\text{-}COOH \xrightarrow{NH_3 \cdot H_2O} p\text{-}H_2N\text{-}C_6H_4\text{-}COONH_4 \xrightarrow{CH_3COOH} p\text{-}H_2N\text{-}C_6H_4\text{-}COOH + CH_3COONH_4$$

Esterification:

The product of esterification is white crystal, its melting point is 91～92℃.

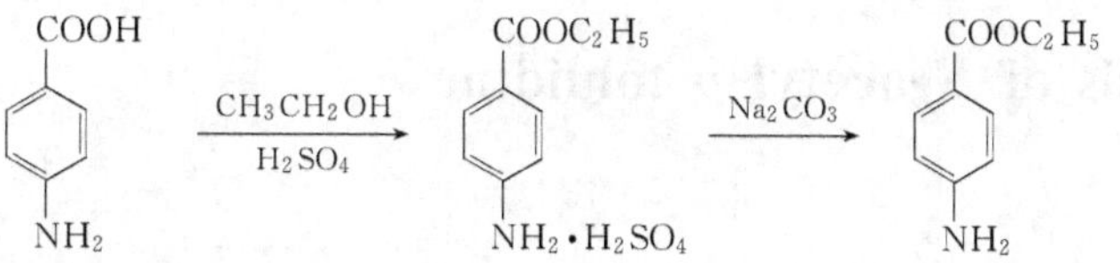

$$\text{COOH-C}_6\text{H}_4\text{-NH}_2 \xrightarrow[H_2SO_4]{CH_3CH_2OH} \text{COOC}_2\text{H}_5\text{-C}_6\text{H}_4\text{-NH}_2 \cdot H_2SO_4 \xrightarrow{Na_2CO_3} \text{COOC}_2\text{H}_5\text{-C}_6\text{H}_4\text{-NH}_2$$

3. 23. 3 Apparatus and Reagents

(1) Apparatus

Three-neck round-bottom flask (150mL), round-bottom flask (150mL), addition funnel (150mL), condenser, glass watch, beaker, magnetic stirrer, Buchner funnel.

(2) Reagents:

p-Nitrobenzoic acid, HCl, Sn, $NH_3 \cdot H_2O$, Na_2CO_3, H_2SO_4.

3. 23. 4 Procedure

3. 23. 4. 1 Reduction

① Arranging assembly Figure 3. 4. Place 4g *p*-nitrobenzoic acid, 9g Sn and stirrer into the three-neck flask, place 20mL HCl (12mol/L) into the addition funnel, then began to react, to drop HCl, after 20～30min, stop reaction.

② Cool the product, then, pour the product into beaker, dropping $NH_3 \cdot H_2O$ (14mol/L) when room temperature while stirring. Filter and wash precipitation by suction on a Buchner funnel, put together the washing solution and filtrate ($V<55$mL).

③ The white crystals are formed when dropping ice-acetic acid into filtrate, further drop a few of acetic acid until the acidity of filtrate, cool, dry and weight.

p-Aminobenzoic acid is yellow crystal, its melting point is 184～186℃.

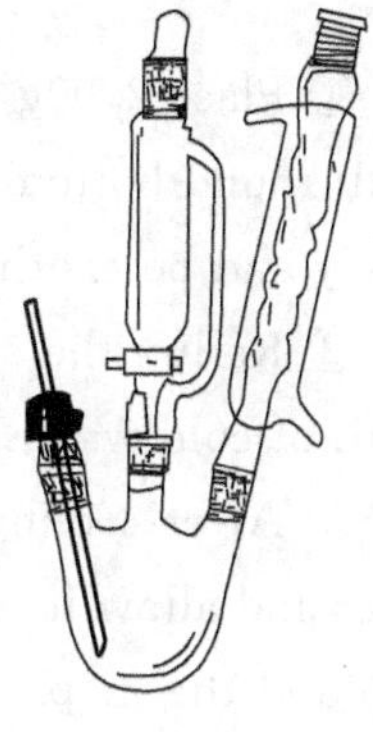

Figure 3. 4 Apparatus for the Synthesis of Benzocaine

3. 23. 4. 2 Esterification

① Arranging assembly of reflux. Place 2g *p*-aminobenzoic acid, 20mL dry ethanol and 2mL H_2SO_4 (18mol/L), boiling chipper into flask. Then pour liquid of reaction into beaker contain 85mL water after reflux 1hour.

② Slowly add to $NaHCO_3$ (s) after cooling while stirring.

③ Slowly dropping liquid of $NaHCO_3$ into when appearing white precipitate. Adjust pH of solution is about 9. Filter and collect product by suction on Buchner funnel, dry and weight.

Pure ethyl *p*-aminobenzoate is white needle crystal, its melting point is 91～92℃.

3. 23. 5 Questions

① How do you judge the end of reduction? Why?

② Please raise other methods of the synthesis of benzocaine. Discuss and compare with them.

3.24 Synthesis of *N*-acetyl-*p*-toluidine

3.24.1 Purpose

Master the Principle and method of preparing *N*-acetyl-*p*-toluidine.

3.24.2 Principle

$$H_3C-C_6H_4-NH_2 + (CH_3CO)_2O \longrightarrow H_3C-C_6H_4-NHCOCH_3$$

3.24.3 Apparatus and Reagents

(1) Apparatus

Suction filtration, reflux apparatus[1].

(2) Reagents

p-Toluidine, acetic anhydride.

3.24.4 Procedure

① Place 2.00g *p*-toluidine, 2.4mL of acetic anhydride[2] and several boiling stones in a 10mL round-bottom flask provided with a reflux condenser. The reaction should start immediately and be exothermic, and the entire solid is dissolved.

② Reflux the solution for 10 minutes. Pour the hot reaction mixture with stirring into 50mL of cold water. a light yellow solid should appear at this point.

③ After cooling thoroughly, filter the solid by suction, wash it three times with cold water and allow it to dry. The weight of the crude product is 2.60g, the percentage yield is 93% and the m.p. is 147～149℃. The crude *N*-acetyl-*p*-toluidine is recrystallized from the mixing solvent of ethanol and water.

Notes

[1] All apparatus must be dry in this experiment.

[2] Acetic anhydride should be redistilled before use. Collect the fraction boiling at 138～139℃.

3.25 Synthesis of Benzoin

3.25.1 Purpose

Master the principle and operation of preparing benzoin from benzaldehyde.

3.25.2 Principle

In this experiment, a benzoin condensation of benzaldehyde is carried out with a biological coenzyme, thiamine hydrochloride, as the catalyst.

Reaction

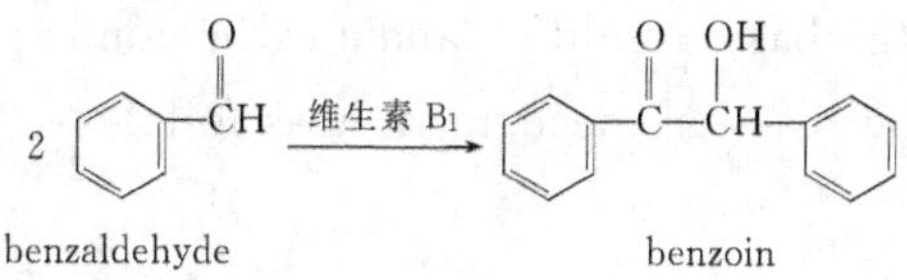

2 benzaldehyde —维生素 B1→ benzoin

3. 25. 3 Apparatus and Reagents

(1) Apparatus

Suction filtration, recrystallization.

(2) Reagents

Benzaldehyde, Vitamin B_1, ethanol, sodium hydroxide.

3. 25. 4 Procedure

① Dissolve 0. 30g of thiamine hydrochloride (Vitamin B_1) in 1. 0mL of water in a 30mL flask. When all of the VB_1 has dissolved, add 3. 0mL of 95% ethanol.

② Stopper the flask and cool the resulting solution with an ice-water bath[1]。

③ Slowly add 1. 0mL of cold 2. 5mol/L sodium hydroxide to the flask, and make pH of the solution is about 10~11[2].

④ Rapidly add 1. 5mL of benzaldehyde to the reaction mixture and sufficiently mix the solution. Stopper the flask and allow it to stand at room temperature for one day. At the end of to reaction period, the benzoin should have separated as fine while crystals.

⑤ When the crystallization has completed, collect the crude product by suction filtration with a Buchner funnel and wash it with a small amount of ice-cold water. Press the crystals as dry as possible and spread them on a fresh filter paper to dry in air.

⑥ Recrystallize the product from 95% ethanol. The weight of benzoin is 0. 6g, and the percentage yield is 38. 5%. The pure product is a white needle-shape crystal[3], m. p. 134~136℃.

Notes

[1] Vitamin B_1 (thiamine) exists in the form of thiamine hydrochloride. It is stable in the acidic condition, but it absorbs water easily, and it is a heat-sensitive reagent, the thiamine in aqueous solution is oxidated easily by oxygen in air. The rate of oxidation may be accelerated by light and some ions such as cupric ion, ironic ion and manganic ion. It should be stored in a refrigerator. Since the thiazole ring is broken easily in basic solution, both the aqueous solutions of thiamine hydrochloride and sodium hydroxide should be cooled thoroughly with an ice-water bath before use.

Vitamin B_1 is a coenzyme. It may replace the extremely toxic sodium cyanide, as the catalyst, in benzoin condensation. The structure of Vitamin B_1 is as follows:

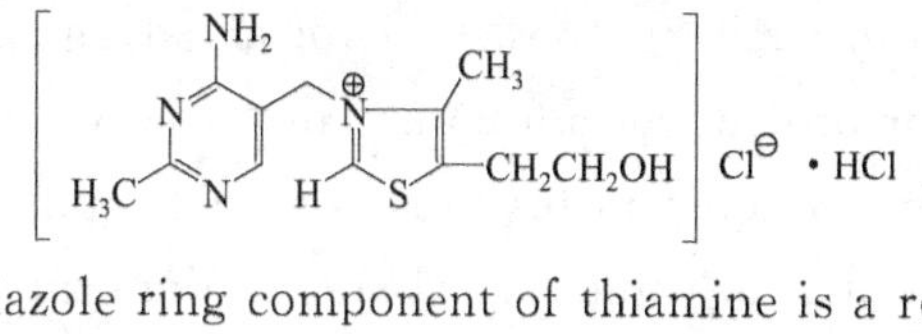

NH_2 CH_3 N H_3C N H S CH_2CH_2OH $Cl^{\ominus}$ · HCl

The proton on the thiazole ring component of thiamine is a relatively acidic proton. The proton is easily removed in the basic condition and a carbanion is produced. So these catalyze the formation of benzoin. The reaction mechanism is as follows:

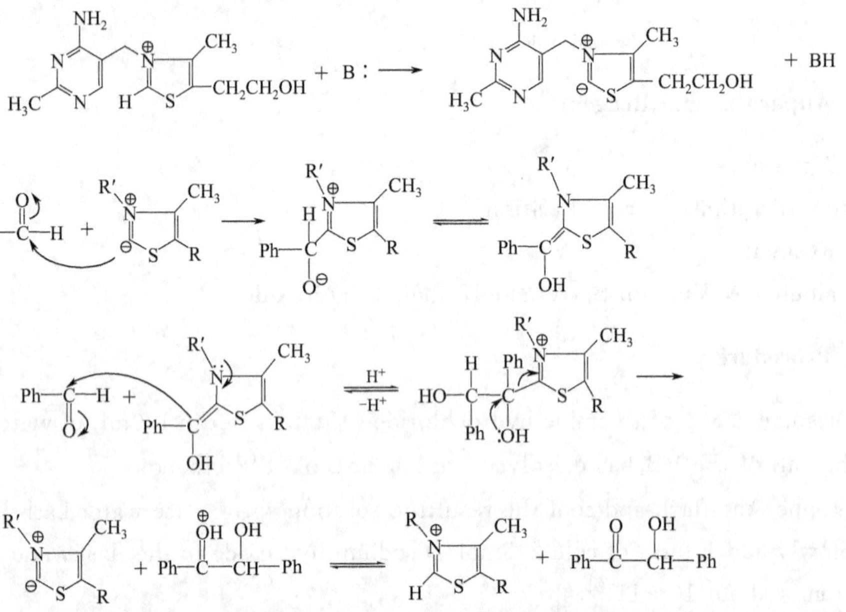

[2] The control of the pH is the key to the benzoin condensation of benzaldehyde. So the benzaldehyde used for this experiment must be free of benzoic acid. The benzaldehyde must be redistilled before use.

[3] Benzoin is a perfumery. The DL-type is a hexagon monoclinic thombic crystal. Both D- type and L-type are needle-shape crystals.

3.25.5 Questions

① How does Vitamin B_1 catalyze the reaction of benzoin condensation?

② What is the function of the sodium hydroxide in this experiment? What is the theoretical amount of it?

③ Interpret the absorbing peaks in the spectrum of benzoin.

3.26 Synthesis of 2-Nitroresorcinol

3.26.1 Purpose

① Learn the application of electrophilic substitution reaction and sulfonation of benzene

② Master the preparing method of 2-Nitroresorcinol.

3.26.2 Principle

In the course of certain organic synthesis it is often advantageous to introduce a group that will effectively block or protect certain potentially reactive sites on a molecule from attack by some specific reagent. Important features of such a group are that it should be easily introduced and easily removed after some crucial step in the synthesis has been performed.

In the process of converting resorcinol (1,3-dihydroxybenzene) to 2-nitroresorcinol, the starting compound is first sulfonated to give resorcinol-4,6-disulfonic acid in which two of the three positions most susceptible to nitration are now blocked. Nitration followed by steam distillation of an acidic solution of the nitrated disulfonic acid to remove the sulfonic acid groups results in pure 2-nitroresorcinol.

Resorcinol $\xrightarrow{2H_2SO_4}$ (resorcinol-4,6-disulfonic acid) $\xrightarrow[HNO_3]{0\sim20℃}$ (2-nitroresorcinol-4,6-disulfonic acid) $\xrightarrow[H^+]{\text{Steam dist.}}$ 2-Nitroresorcinol

3.26.3 Apparatus and Reagents

(1) Apparatus

Beaker, addition funnel, steam distillation, recrystallization apparatus.

(2) Reagents

Powder resorcinol, concentrated sulfuric acid, nitric acid, concentrated sulfuric acid, 95% ethanol.

3.26.4 Procedure

① Place 7.7g (0.07mol) of powered resorcinol[1] in a 150mL beaker and add 28mL (50.4g, 0.515mol) of concentrated sulfuric acid (98%, d = 1.84). If a thick slurry of the 4,6-disulfonic acid does not form in a few minutes, warm the mixture to 60～65℃. Allow the slurry to stand for 15min.

② Prepare a mixture of 4.4mL (4.38g, 0.0693mol) of nitric acid (70%～72%, d = 1.42) and 6.2mL (11.9g, 0.116mol) of concentrated sulfuric acid, and cool it in an ice bath.

③ Cool the slurry in an ice-salt bath to a temperature of 5～10℃, stir it, and slowly (dropwise) add the cold acid solution from a addition funnel suspended over the beaker. The temperature of the reaction mixture should not exceed 20℃.

④ After the yellowish mixture has stood for 15min, it should be cautiously diluted with 20g of crushed ice so that the temperature never exceeds 50℃.

⑤ Transfer the mixture to a 500ml round-bottom flask, add 0.1g of urea[2].

⑥ And carry out an indirect steam distillation until no more of the orange-red, solid 2-nitroresorcinol appears in the condenser or until yellow needles of the undesired 4,6-dinitro-re-sorcinol (m.p. 215℃) appear in the condenser. The product will usually appear after about 5min of steam distillation. The product may not steam-distill if too much steam has condensed in the distillation flask. In this event shut off the steam and heat the flask (a Bunsen burner or a heating mantle will be required) until sufficient water is removed to increase the sulfuric acid concentration in the flask to the point at which desulfonation will occur and product will again distill. If the condenser becomes filled with solidified product, turn off the cooling water for a few minutes until the product has melted and flowed into the receiver.

⑦ Cool the distillate in an ice bath and filter it with suction.

⑧ Recrystallize the product from dilute aqueous ethanol by first dissolving it in 95% ethanol (ca. 3mL per gram of product), filtering it hot, adding water slowly until cloudiness (or small amounts of precipitate) persists, and allowing the solution to cool slowly.

The yield of 2-nitroresorcinol is 2.5～3.5g. The melting point of pure 2-nitroresorcinol is 84～85℃.

Notes

[1] In order to be sulphonated completely, the resorcinol should have been previously ground to fine power in a mortar.

[2] When urea is added, the salt of the urea and the nitric acid is formed. As the salt dissolves in water, the excess of the nitric acid is remove.

3.26.5 Questions

① Why might sulfonation be expected to occur most readily in the 4-and 6-positions rather than the 2-position? How does the overall reactivity (toward electrophilic attack) of the disulfonic acid compare with that of resorcinol?

② Write out a step-by-step mechanism to explain the desulfonation process.

③ Devise a mechanism to account for formation of the 4,6-dinitroresorcinol?

3.27 Synthesis of Quinoline

3.27.1 Purpose

Learn the principle and method of Skraup synthesis of quinoline.

3.27.2 Principle

$$\underset{OH}{CH_2}-\underset{OH}{CH}-\underset{OH}{CH_2} \xrightarrow[-H_2O]{\text{浓 } H_2SO_4} \underset{OH}{CH_2}-CH=\underset{OH}{CH} \rightleftharpoons \underset{OH}{CH_2}-CH_2-\underset{O}{\overset{\|}{C}}-H \xrightarrow{\text{浓 } H_2SO_4} CH_2=CH-\underset{O}{\overset{\|}{C}}-H$$

$$C_6H_5NH_2 + CH_2=CH-CHO \longrightarrow \left[C_6H_5NH-CH_2-CH_2-CH=O \rightleftharpoons C_6H_5NH-CH_2-CH=CH-OH \right]$$

$$\xrightarrow{-H_2O} \text{1,2-dihydroquinoline} \xrightarrow[C_6H_5NO_2]{-H_2} \text{quinoline}$$

Quinoline can be prepared by heating a mixture of aniline, glycerol and sulfuric acid with a weakly oxidizing agent like nitrobenzene. This strategy was called Skraup synthesis. In the Skraup synthesis of quinoline the principal difficulty has always been the violence with which the reaction takes place; it often gets beyond control in the majority of cases. By the addition of ferrous sulfate, the reaction generally proceeds relatively smoothly.

3.27.3 Apparatus and Reagents

(1) Apparatus

Reflux, steam-distillation, extraction, distillation apparatus.

(2) Reagents

Glycerol, aniline, nitrobenzene, con. sulfuric acid, 30% sodium hydroxide, ferrous sulfate, sodium nitrite, ether, solid sodium hydroxide, starch-potassium iodide paper.

3.27.4 Procedure

① In a 500mL round-bottom flask, fitted with an efficient reflux condenser, are placed, in the following order, 4g of powdered crystalline ferrous sulfate[1], 29.9mL of glycerol[2], 9.3mL of aniline and 6.7mL of nitrobenzene. After mixing well, 18mL concentrated sulfuric acid is added slowly with shaking-up [3].

② The mixture heated gently over a free flame. As soon as the liquid begins to boil, the flame is removed, since the heat evolved by the reaction is sufficient to keep the mixture boiling for one-half to one hour.

③ When the boiling has ceased the heat is again applied and the mixture boiled for two and one-half hours [4].

④ It is then allowed to cool to about 100℃ and the flask is connected with the steam-distillation apparatus, Steam is passed in until no further droplets of oil can be seen. This removes all the unchanged nitrobenzene.

⑤ The current of steam is then interrupted, the receiver is changed and 30% sodium hydroxide solution is added cautiously to neutralize the sulfuric acid until the aqueous solution is alkaline [5].

⑥ Steam is then passed again in as rapidly as possible until all the quinoline and unchanged aniline have distilled. The distillate is acidified with concentrated sulfuric acid until the oily material is dissolved absolutely.

⑦ The solution is cooled to about 5℃ and a saturated solution of sodium nitrite (3g bodium nitrite and 10mL water) is added until one drop of solution change the starch-potassium iodide paper to blue[6].

⑧ The mixture is then warmed on a steam bath for 15 minutes or until active evolution of gas ceases[7].

⑨ After cooling, the mixture is basified with 30% sodium hydroxide solution and is then distilled with steam.

⑩ Oil phase is separated from the distillate and water phase is extracted with ether twice

(25ml each time). Collect the oil phase and extraction and dried over sodium hydroxide overnight.

⑪ After recycling ether, the residue is then distilled and collect the fraction which boils at 234～238℃[8] yielding 8～10g[9]. (Lit. b. p. 238℃; 114℃/17mmHg).

Notes

[1] In the Skraup synthesis of quinoline the principal difficulty has always been the violence with which the reaction generally takes place, it gets beyond control in the majority of cases. By the addition of ferrous sulfate, which appears to function as an oxygen carrier, the reaction is avoided too violent.

[2] In a number of experiments, the glycerol used contained an appreciable amount of water. Under these conditions, the yield of product is much lower. To get rid of water, the glycerol can be heated in evaporating dish at 180℃ in the hood. After cooling to 100℃, the glycerol is placed in desiccators with sulfuric acid as desiccant for further use.

[3] It is important that the materials should be added in the correct order; should the sulfuric acid be added before the ferrous sulfate, the reaction may start at once.

[4] It is also important to mix the materials well before applying heat; the aniline sulfate should have dissolved almost completely, and the ferrous sulfate should be distributed throughout the solution. To avoid danger of overheating, it is well to apply the flame away from the center of the flask where any solids would be liable to congregate.

[5] When acidified or basified, the reaction solution should be cooled appreciably and stirred thoroughly. After that, determine the solution whether it is shown acidity or basicity.

[6] It proceeds very slowly when the diazotization reaction is close to completion. 2～3 minutes later after sodium nitrite solution is added, it would be determined the existence of nitrous acid.

[7] The properties of diazotization reaction and diazonium salt are applied in this experiment to get rid of the aniline in the quinoline product. Following shows the procedure of the reaction.

[8] It is suggested that vacuum distillation is applied and collect the fraction at 110～114℃/14mmHg, 118～120℃/20mmHg or 130～132℃/40mmHg. The product is colorless, transparent liquid.

[9] The percentage yields have been based on the amount of aniline taken. It would probably be more legitimate to base the calculation on the amounts of aniline taken and of nitrobenzene not recovered, since undoubtedly the latter is reduced to aniline during the course of the reaction.

3.27.5 Questions

① Steam distillation is used for three times in this experiment, please answer the following questions

a. Are there any aniline and quinoline in the distillate of the first steam distillation? Why?

b. Why should the 30% sodium hydroxide be added to neutralize the acid produced by the reaction until the aqueous solution is alkaline before the second or third steam distillation?

c. What is the convenient way to check if there is aniline in the distillate of the second steam distillation?

② What are the expected products if using *p*-methylaniline or *o*-methylaniline rather than aniline in the Skraup synthesis? How to choose nitro compound?

③ Explain the main factors which influence the quality and yield of the product.

3.28 Synthesis of Nikethamide

3.28.1 Purpose

To know the principle and method of preparing nikethamide from carboxylic acid and amine.

3.28.2 Principle

Nikethamide (*N*, *N*-diethyl nicotinic amide) can be prepared when nicotinic acid reacts with diethylamine. The equation is as follow:

$$\text{(N-ring)}\text{COOH} + HN(C_2H_5)_2 \longrightarrow \text{(N-ring)}\text{COOH} \cdot HN(C_2H_5)_2$$

$$\text{(N-ring)}\text{COOH} \cdot HN(C_2H_5)_2 \xrightarrow{POCl_3} \text{(N-ring)}\cdot HCl\text{—C(=O)}N(C_2H_5)_2 + H_3PO_4$$

$$\text{(N-ring)}\cdot HCl\text{—C(=O)}N(C_2H_5)_2 + NaOH \longrightarrow \text{(N-ring)}\text{—C(=O)}N(C_2H_5)_2 + NaCl + H_2O$$

3.28.3 Apparatus and Reagents

(1) Apparatus

Three-necked round-bottom flask, separatory funnel, extraction, distillation apparatus.

(2) Reagents

Nicotinic acid, diethylamine, phosphorus oxychloride, 10% potassium permanganate solution.

3.28.4 Procedure

① In a 100mL dry three-necked round-bottom flask, are placed 12.3g of nicotinic acid, 10.2g of diethylamine [1]. Heat the flask slowly with swirling, until all the solid has been dissolved [2].

② After cooling under 60℃, 8.4g of phosphorus oxychloride[3] is added dropwise to the flask at such a rate that keep the temperature is below 140℃. Then maintain the temperature at 135℃ for 2.5 hours.

③ After cooling the mixture to 80℃, 12mL of water is slowly added. When the temperature is below 55℃, add 20% sodium hydroxide solution until the pH is 6～7[4]. Transfer the mixture into a separatory funnel, remove the aqueous layer.

④ And place the organic layer to a 100mL flask. Then dilute it with 10mL of water, add 3mL of 10% potassium permanganate solution, and shake it. Decolorize and filter the oxidized.

⑤ Mixture with a funnel covered with decolorizing carbon (about 3g). Wash the filter cake with little water. Combine the washing liquor with the filtrate. Then add 10% potassium carbonate solution to make the pH7.5. Transfer the solution to a separatory funnel. Extract it with chloroform for 4 times (20mL twice and 15mL twice), combine the chloroform, wash it with distilled water [5] for 4 times (8mL each), and dry it over anhydrous calcium carbonate.

⑥ The chloroform is removed by distillation under ordinary pressure and the residue is distilled under reduced pressure collecting the pale yellow distillate at 160～170℃/10～15mmHg, the yield of *N*, *N*-diethyl nicotinic amide is about 12.5g (Lit. m. p. 24～26℃, b. p. 175℃/25mmHg: 158～159℃/10mmHg; 128～129℃/3mmHg).

Notes

[1] Diethylamine and phosphorus oxychloride should be redistilled before use, and nicotinic acid should be dried under 80℃.

[2] If the solid dissolve, heating is left out.

[3] Phosphorus oxychloride liberates hydrogen chloride upon absorbing moisture. Keep it in a dry condition, and it had better be distilled in a hood.

[4] Keep the temperature below 60℃ during neutralization.

[5] Nikethamide is a drug. If it is washed with tap water, some impurities would be introduced.

3.28.5 Questions

① What is the role of phosphorus oxychloride in the formation of amide?

② What would happen if temperature is higher than 60℃ during the neutralization by sodium hydroxide?

③ Why do we use 10% potassium permanganate solution to wash the oil layer?

3.29 The Reduction of Camphor

3.29.1 Purpose

① Learn the principle and operation method of reduction of camphor by sodium borohydride.

② To know application of thin-layer chromatography.

3. 29. 2 Principle

Camphor will be reduced to a mixture of bomeol and isobomeol which are diastereoisomers by the action of sodium borohydride. Because of the high stereoselectivity, the main product is isobomeol. Bomeol and isobomeol have different physical property and polarity.

$$\text{camphor} \xrightarrow{NaBH_4} \text{bomeol} + \text{isobomeol}$$

camphor bomeol isobomeol

3. 29. 3 Apparatus and Reagents

(1) Apparatus

Flask, recrystallization, distillation apparatus.

(2) Reagents

Camphor, sodium borohydride, methanol, ether, anhydrous sodium or magnesium sulfate.

3. 29. 4 Procedure

Place the camphor (1. 0g) in the 25mL flask, add methanol (10mL). Stir with a glass rod or microspatula until the solid has dissolved. In portions, cautiously and intermittently add 0. 6g of sodium borohydride[1] to the solution at room temperature. If necessary, cool the flask in an ice bath to control the temperature of the reaction mixture. When all the borohydride is added, heat the contents of the flask to reflux until all the sodium borohydride has dissolved.

Allow the reaction mixture to cool to room temperature, and then carefully add to ice-water (20mL) while stirring. Collect the white solid which forms by filtering through a Buchner funnel, and suction-dry it for several minutes while you clean and dry the 100mL flask. Transfer the solid back to the clean flask, and add 25mL of ether [2] to dissolve the solid followed by 6～7 microspatulas of anhydrous sodium or magnesium sulfate. After about 5 minutes of drying, transfer the solution (leaving behind the drying agent) to a previously weighed beaker or flask. Evaporate the solvent off in the hood to obtain the product as a white solid. Recrystallize the crude product from aqueous ethanol. The yield of the product is about 0. 6g.

Weight the product and calculate the percentage yield. Determine the melting point (literature melting point: isoborneol 212℃) in a sealed capillary tube and record the IR spectrum as a KBr pellet.

Notes

[1] Sodium borohydride liberates hydrogen upon reaction with water; so it should be kept it in a drying condition

[2] Ether is highly flammable, so keep it away from fire.

3.29.5 Questions

① What should we pay attention to while determining the melting point of the product?

② What kind of method we can use to distinguish and identify bomeol and isobomeolex-cept to using IR spectroscopy?

③ What is the main product in the reaction of sodium borohydride with ortho-camphor?

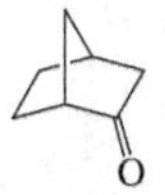

O

ortho-camphor

3.30 Synthesis of 3-(2,5-xylyloxyl)Propyl Chloride and Monitoration of Reaction Process

3.30.1 Purpose

① Learn the principle and operation method of preparing 3-(2,5-xylyloxyl)propyl chloride

② To know application of thin-layer chromatography.

3.30.2 Principle

3-(2,5-Xylyloxyl) propyl chloride was synthesized from 2,5-dimethylphenol reacting with 1-bromo-3-chloropropane in the presence of sodium hydroxide. Because 2,5-dimethyl-phenol and 1-bromo-3-chloropropane exist respectively in organic phase and aqueous phase, the time of reaction is long and the yield is low. If the reaction is catalyzed by phase-transfer catalyst (PTC, for example TEBA), the time of reaction is shortened and the yield is increased by 30%or more.

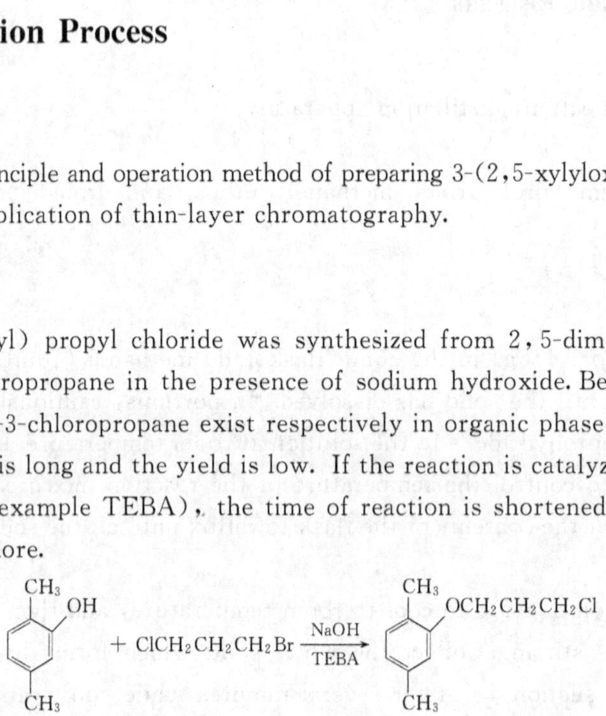

$$CH_3,\ OH,\ CH_3 \text{(2,5-dimethylphenol)} + ClCH_2CH_2CH_2Br \xrightarrow[TEBA]{NaOH} CH_3,\ OCH_2CH_2CH_2Cl,\ CH_3$$

3.30.3 Apparatus and Reagents

(1) Apparatus

Four-neck round-bottom flask, mechanical stirrer, dropping funnel, Extraction/ separatory, distillation apparatus.

(2) Reagents

2,5-Dimethylphenol, l-bromo-3-chloropropane, sodium hydroxide, TEBA, ethyl ether, saturated sodium chloride solution, anhydrous magnesium sulfate, benzene, petroleum (60~90℃).

3.30.4 Procedure

① In a 100mL four-neck round-bottom flask provided with a mechanical stirrer, thermometer, dropping funnel, and reflux condenser are placed 3.8g (0.031mol) 2,5-dimethylphenol, 7.9g (0.05mol) 1-bromo-3-chloropropane, 0.2g TEBA. The mixture is stirred vigorously [1] and 30mL of 1.6mol/L sodium hydroxide is added from a dropping funnel at such a rate that the whole is added in about one hour, while the temperature is maintained between 90~94℃. Stirring is continued at about 100℃ for another four hours until the pH is 6~7[2]. Cool the mixture to room temperature, transfer the mixture to a separatory funnel, and separate the organic liquid from the aqueous layer. Extract the aqueous layer with diethyl ether (15mL, 10mL×2), combine the organic liquid and the extract liquor, wish it with 10mL of saturated aqueous sodium chloride, and separate the layers carefully. After dried over anhydrous magnesium sulfate, the ether is removed as completely as possible and the residual is distilled under reduced pressure collecting the distillate at 108~110℃/266Pa [3]. The product is light yellow liquid and the yield is about 4.7g (77.7 percent of the theoretical amount).

② Experimental conditions for TLC analysis

eluents: benzene- petroleum = 1 : 1 (volume).

adsorbents: silica gel (GF_{254}).

developing distance: 10cm.

visualization methods: a low-intensity ultraviolet lamp.

Notes

[1] Because the reaction is carrying through catalyzed by phase-transfer catalyst, it is necessary to stir vigorously.

[2] If it is not, add extra base.

[3] In addition to 3-(2,5-xylyloxyl)propyl chloride is obtained, 3-(2,5-xylyloxyl) propyl bromine, b. p. 124~132℃/266Pa, may be produced.

3.30.5 Questions

① Why should 2,5 -dimethylphenol react with 1-bromo-3-chloropropane in the presence of phase-transfer catalyst?

② Suggest methods for preparing the following ethers via Williamson syntheses.

A. $CH_3CH_2OCH_2CH_3$ B. $CH_3CH_2OCH(CH_3)_2$ C. $CH_3CH_2CH_2CH_2OCH_2CH_2CH_3$

③ Why does only little 3-(2,5-xylyloxyl) propyl bromine be produced?

④ Why is it necessary to stir vigorously?

Part 2 Extraction and Separation

3.31 The Recrystallization of Benzoic Acid

3.31.1 Purpose

① To know the principle and method purifying the solid organic compound by recrystallization;

② Master the experiment operation of recrystallization.

3.31.2 Principle

Recrystallization (about principle and procedure see 2.6.1) is the most frequently used operation for purifying organic solids. This technique is based on the fact that the solubility of an organic compound in a given solvent will often increase greatly as the solvent is heated to its boiling point. When it is first heated in such a solvent until it dissolves, then cooled to room temperature or below, an impure organic solid will usually recrystallize from solution in a much purer form than its original form. Most of impurities will either not dissolve in the hot solution (from which they can be filtered), or remain in dissolved form in the cooled solution (from which the pure crystals are filtered).

In this experiment, The water is reagent due to the benzoic acid can hardly dissolve in cold water, and it can be dissolved slightly by hot water.

3.31.3 Apparatus and Reagents

(1) Apparatus

Round-bottom flask (250mL), condenser, beaker, hot filtering funnel, Buchner funnel, filtering flask

(2) Reagents

Benzoic acid, activated carbon, pure water.

3.31.4 Procedure

① Assemble Apparatus Figure 2.14.

② Place 3.0g of impure benzoic acid in a 250mL round-bottom flask, add 80mL of pure water and heat to boiling with the aid of a burner or hot plate. Cool the solution slightly, add about 0.1g of activated carbon, and reheat it to boiling for a few minutes.

③ Filter the hot solution by hot filtering funnel with a fluted filter paper (shown as Figure 2.4), collect the filtrate with a beaker.

④ Set the hot filtrate in place and cool it down to make the crystals precipitate.

⑤ Separate the crystals of Benzoic acid by vacuum filtration (see Figure 2.2).

⑥ Spread the crystals onto a piece of filter paper and allow them to air-dry completely.

⑦ Weight and calculate the percent recovery.

3.31.5 Questions

① How many steps does the process of recrystallization involve? What is the purpose of each step?

② Why should the amount of the solvent consumed in recrystallization be neither too much nor too little? How should you do it exactly?

③ Why should the amount of the solvent added in dissolving the rude product be less than that by calculations?

3.32 Extraction and Purification of Nicotine

3.32.1 Purpose

① Master the operation of steam distillation.

② To be familiar with the principle of nicotinic extraction and purification, and chemical property of nicotine.

3.32.2 Principle

Nicotine mainly existed tobacco leaf. Its boiling point is 247℃. It is toxic. Its structural formula:

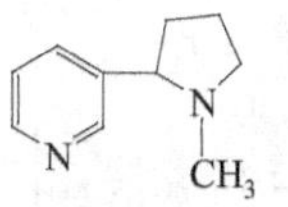

At room temperature, nicotine is colorless and oily liquid. It is stronger alkalinity, it may form its salt when react with acid and dissolved water. In the presence of NaOH, nicotine also extracted from its solution at 100℃, extracted nicotine has appreciable vapor pressure (1.333kPa).

Therefore, the steam distillation is often employed for separating and purifying of nicotine.

Nicotine may react with phenolphthalein indicator and oxidant. Like other alkaloids, nicotine react with alkaloidal reagent to form precipitation.

3.32.3 Apparatus and Reagents

(1) Apparatus

Steam-generator (round-bottom flask), T-shaped glass tube, round-bottom flask, condenser, adapter, flask, glass tube, screw clamp, beaker.

(2) Reagents

Silk tobacco; HCl (3mol/L); NaOH (6mol/L), phenolphthalein, $KMnO_4$ (0.3/L), picric acid, HAc (8mol/L), Na_2CO_3 (0.5mol/L), I_2, KHgI.

3.32.4 Procedure

3.32.4.1 Extraction of Nicotine

Place 2.0g of silk tobacco into 250mL beaker, add 40mL of HCl (3mol/L), heating to boil, during heating, it must continue mixed and added water to keep the level of the liquid. After boiling 20 minutes, cooling and filtering (by Buchner funnel). The filtering solutions are counteracted by NaOH (6mol/L) to alkaline (pH paper test, pH>7).

3.32.4.2 The Steam Distillation—Separation and Purification of Nicotine

① Arrange an apparatus for steam distillation (see Figure 2.9).

② Add water (about 3/4 volume of the generator) and 2 tiny boiling chips in the steam generator, introduce above filtering solution to the three-neck flask.

③ Heating the water in the steam generator to boiling.

④ Clamp the screw clamp when vapor give off from the under end of the T-shaped tube. adjust a heating source, so that a moderately rapid and steady current of steam is passed into the filtering solution.

⑤ Collect 30mL of the distillate.

⑥ When the distillation is finished, open the screw clamp of the drain tube and disconnect the distillation flask before shutting off the steam.

3.32.4.3 Test of Chemical Properties of Nicotine

(1) Alkalinity

The steam distillate 1mL, is placed in one tube, add 1 drop phenolphthalein indicator to observe the phenomenon.

(2) Oxidation

The steam distillate 5mL, is placed in one tube, add 1 drop $KMnO_4$ (0.3mol/L) and 3 drops Na_2CO_3 (0.5mol/L), oscillate the tube, to observe the color change.

(3) Reaction of precipitation

① The steam distillate 5mL, is placed in one tube, add 5 drops the solution of picric acid (slowly dropping), to oscillate the tube and observe the phenomenon.

② The steam distillate 5mL is placed in one tube, add 5 drops the solution of I_2. to oscillate the tube and observe the phenomenon.

③ The steam distillate 5mL, is placed in one tube, add 1 drops HAc and slowly add the solution KHgI, to oscillate the tube and observe the phenomenon.

3.32.5 Questions

① Please describe the basic principle of steam distillation.

② What are the functions of the T-shaped tube and safety column?

③ How do you judge that steam distillation is complete?

3.33 Isolation of Caffeine From Tea Leave

3.33.1 Purpose

① Learn two different methods of extraction including main principles and basic operations.

② To know the process of sublimation.

3.33.2 Principle

Caffeine can be isolated from tea by means of microwave-assisted extraction as well as traditional Soxhlet extraction.

Caffeine belongs to a group of alkaloid compounds called the xanthines. Its chemical name is 1,3,7-trimethyl-2,6-dioxopurine and it is derivative of purine as shown below:

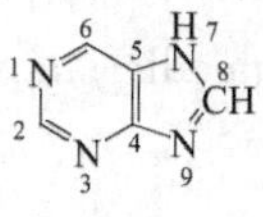

Purine

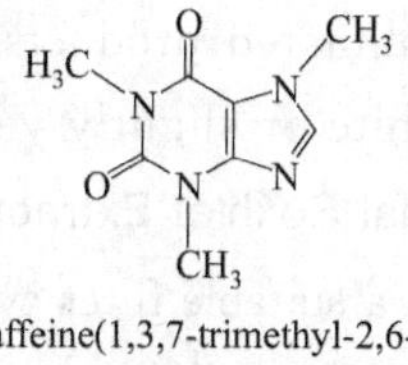

Caffeine(1,3,7-trimethyl-2,6-dioxoppurine)

Hydrated caffeine is fairly soluble in 80℃ hot water, or acetone, or ethanol, or chloroform and so on. It can lose crystal water at 100℃, and start sublimating. The process of sublimation is very obviously at 120℃, the sublimation can go fast when the temperature is above 178℃. The melting point of the dehydrated caffeine is 234.5℃.

3.33.3 Apparatus and Reagents

(1) Apparatus

Microwave, Soxhlet extractor, round-bottom flask (150mL), condenser, glass watch, adapter, funnel, thermometer, beaker, evaporating dish.

(2) Reagents

Tea leaves, ethanol (95%), CaO powder.

3.33.4 Procedures

3.33.4.1 The Method of Microwave-assisted Extraction[1]

① Place 10g tea leaves into a 250mL iodimetric flask. Add 120mL of ethanol (95%) as well as some boiling stones.

② Put the iodimetric flask into a home microwave, adjust microwave power up to 320W, radiate for 50～60s[2], and then take the flask out and cool it down.

③ Repeat the above procedures three or four times[3], and then filter to remove the tea leaves from the ethanol solution.

④ Place the filtrate in a distillation apparatus with a water bath and distill out most of the solvents, which can be reused afterwards, until the residual liquid in the flask is 5～8mL.

⑤ Pour the residue into an evaporating dish, wash the flask with small amounts of distilled ethanol three times, and then add them to the residue in the evaporating dish together.

⑥ Add 2.5g CaO powder[4] into evaporating dish and stir the mixture continuously. Then put it on a boiling water bath and evaporate to dryness.

⑦ Move the evaporating dish onto the asbestos pad, and heat the crude caffeine with a low, moderate flame.

⑧ Pierce many small holes in a large round filter paper and cover over an appropriately sized glass funnel with it. Put a small amount of cotton batting into the neck of the funnel and place on the evaporating dish.

⑨ Sublime the caffeine[5] with a small flame[6] at first, and then stop heating as soon as some brown oil substances appear on the wall of glass funnel. Cool and collect the caffeine crystals on the filter paper.

⑩ Re-sublime the residue of caffeine twice with a slightly larger flame after stirring completely. Afterwards put all sublimated products together. Weigh and calculate the yield (70 to 80mg). Pure caffeine is a white or slightly yellow needle-shaped crystal, m. p. 238℃.

3.33.4.2 The Method of Traditional Soxhlet Extraction

① Place 10g of tea leaves[7] into a suitable filter cylinder, then put it into a Soxhlet extractor.

② Add 60mL of 95% ethanol and put boiling stones into a round-bottom flask. Prepare a reflux apparatus with a water bath (see Figure 2.6).

③ Reflux continuously and extract for two hours until the color of extracted liquid is very light. Stop heating instantly[8] once the liquid just siphons back to the flask from the Soxhlet extractor.

④ The rest of the steps will follow the procedures mentioned in microwave-assisted extraction[9] from No. ③ to No. ⑨.

Notes

[1] Microwave-assisted extraction can save more than two hours compared with other methods of extraction.

[2] Do not boil vigorously or boil to dryness during microwave heating.

[3] Cool down before repeating microwave heating.

[4] Neutralize with CaO to remove acidic substances such as tannic acid.

[5] If the mixture is not completely dry, there exists small water drops on the wall of glass funnel at the beginning of sublimation. If so, move the flame away, wipe water drops off, and continue with sublimation.

[6] Controlling temperature is the key step during the process of sublimation which has a direct influence on the quality and yield of caffeine. Too high a temperature results in a charred substance or makes caffeine yellow.

[7] The filter paper must be tightly close to the wall of Soxhlet extractor, Its height is not higher than the one of siphon arm with Soxhlet extractor.

[8] Control the reflux speed, usually 8～10time/2h.

[9] The yield of caffeine is 30～40mg.

3.33.5 Questions

① Assume that your crude caffeine is nearly pure caffeine. How does the percentage yield of crude caffeine obtained in this experiment compare with the expected value of 3% caffeine in tea leaves? Is there a discrepancy between your result and the normal value? Give a feasible explanation.

② Try to make a comparison between microwave-assisted and Soxhlet extraction and describe their features.

3.34 Isolation of Effective Components from the Citrus

3.34.1 Purpose

① To master the method and principle of isolation of effective components from the plants.

② Further review the basic operation technique of the steam distillation and extraction.

3.34.2 Principle

Lemon oil is a kind of essential oils which exist in the fresh seedcase of lemon and orange etc. The seedcase contains 0.35% oil, yellow liquid, has a strong lemon aroma ρ=0.857~0.862 (15/4℃). n_D^{20} = 1.474~1.476. $[\alpha]_D^{20}$ = +57°~+61°. The main component of the oil is limonene, its content reaches 80%~90%. The main aroma consists of 3%~5.5% citral and pinene, etc. They were used to prepare drink, soap, cosmetic and essence.

Utilizing the method of stream distillation, we can distillate essential oils and vapor together from the grinding orange peel. Then we can get the lemon oil after the extraction by organic solvent and distilling the solvent.

3.34.3 Apparatus and reagents

(1) Apparatus

Three-neck round-bottom flask (500mL), condenser, adapter, Erlenmeyer flask (50mL, 100mL, 250mL), funnel (125mL), round-bottom flask (50mL), still head, thermometer (100℃), beaker (1000mL), steam generator.

(2) reagents

Orange peel (fresh) 50g, petroleum ether, anhydrous sodium sulfate.

3.34.4 Procedure

3.34.4.1 Steam Distillation

① Assemble the steam distillation apparatus (Figure2.9).

② Put the shatter of 50g fresh orange peel into a 500mL of three-neck round-bottom flask, adding about 250mL water[1].

③ Begin to steam distillation, control the speed of distilling is 2~3drops/s. The distillation is over when collect the distillate about 80mL[2].

3.34.4.2 Extraction and Drying

① Putting the distillate into separating funnel.

② The distillate are extracted three times by 30mL petroleum ether.

③ The upper-liquor was collected, and all extract liquid was combined.

④ Putting them into 50mL dry Erlenmeyer flask, adding proper anhydrous sodium sulfate, shaking it until the liquid is clear.

3.34.4.3 Concentration

① The dry extract liquid are putted into 50mL dry round-bottom flask.

② Assemble the distillation apparatus for distilling low boiling point compounds, heating on the water bath (<100℃).

3.34.4.4 Collect the product

Then you can get a little of yellow oil after petroleum ether was recovered[3].

3.34.4.5 Determination

The purity of product is determined by refractive index or optical rotation.

Notes

[1] Seedcase should be cut as shatter as possible, they are directly cut into the beaker to prevent the essential oils were loss.

[2] At that time, some oil drops maybe still existed in the distillate. But its content was less, and you'd better stop distilling.

[3] Petroleum ether was also recovered by rotary evaporator.

3.34.5 Questions

① Why the method of steam distillation is adopted to distill the essential oils.

② What is the reason for the yield of oil decreased when the dry orange peel is used to extract Lemon Oil?

③ According to the experiment, Searching some relative data, trying to advance 1～2 experimental schemes about distilling plant essential oil, please.

3.35 Isolation of Benberine from *Coptis chinensis* Franch

3.35.1 Purpose

① To master the method and principle of isolation of benberine from *Coptis chinensis* Franch.

② Further review the basic operation technique.

3.35.2 Principle

Benberine is yellow crystal, m. p is 145℃, dissolved in hot enthanol and hot water, but it is insoluble for ether. Its salt is very purified.

The purity of benberine can be determinated by melting point, TLC and UV-spectrum. The structure of benberine is shown:

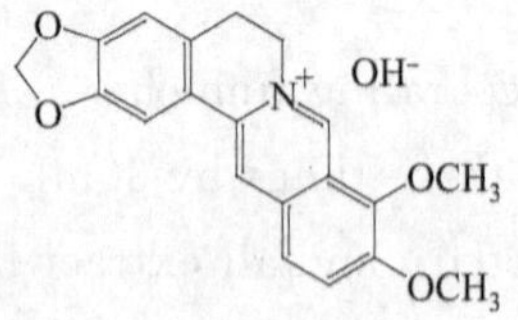

OH^- OCH_3 OCH_3

3.35.3 Apparatus and reagents

(1) Apparatus

Round-bottom flask, condenser, beaker, Soxhlet extractor glass watch, adapter, funnel, thermometer,

(2) Reagents

Coptis chinensis Franch (powder), acetone, chloroform, methanol ethanol, HCl, acetic acid, calcareous milk, Al_2O_3 thin plate.

3.35.4 Procedure

3.35.4.1 Extraction

① Place 10g of *Coptis chinensis* Franch (powder) into a suitable filter cylinder, then put it into a Soxhlet extractor.

② Add 100mL of ethanol (95%) and put boiling stones into a round-bottom flask. Prepare a reflux apparatus with a water bath (see Figure 2.6).

③ Reflux continuously and extract for two hours until the color of extracted liquid is very light. Stop heating instantly once the liquid just siphons back to the flask from the Soxhlet extractor.

④ Place the filtrate in a distillation apparatus with a water bath and distill out most of the solvents, which can be reused afterwards, until the residual liquid in the flask is sirup.

3.35.4.2 Purification

① Add 30~40mL of acetic acid (0.2mol/L) into the solution of sirup, Heating until it dissolved.

② Filter by suction.

③ Dropping HCl (12mol/L) into the filtrate until mud is formed (about 10mL).

④ Cooling with ice-water, can get the salt of benberine (yellow needle crystal).

⑤ Filter by suction, washing crystal with cool water twice and acetone. Drying product.

⑥ Add hot water into the products until they are dissolved, boiling and adjusting pH= 8.5~9.8 with calcareous milk, cool and filter.

⑦ Continuous to cool at room temperature, washing crystal with cool water twice and acetone.

⑧ Drying product at 50~60℃. Weight.

3.35.4.3 Determination

(1) Determination of Melting Point

It is m. p. is 145℃.

(2) TLC Analysis

Operation: see "2.7 Chromatographic Techniques".

Sample: the ethanol solution of benberine.

Developing solvent: chloroform-methanol 9 : 1 (*V/V*).

3.35.5 Questions

① Which kinds of alkaloid is the benberine?

② In this experiment, why do you adjust pH with calcareous milk?

3.36 Paper Chromatography of Amino Acid

3.36.1 Purpose

① To know the principle of paper chromatography.

② Learn the method of separating amino acids.

3.36.2 Principle

When a small spot containing a mixture of amino acids is placed near the bottom of a piece of filter paper, and the filter paper is placed in a covered jar containing a small amount of suitable solvent, the solvent moves up the filter paper, carrying each different amino acid in the mixture up the paper to a different extent. This results in a series of spots on the paper, each spot corresponding to a different compound. This is the steps of paper chromatography; and is showed in Figure 2.22, Figure 2.23.

The water absorbed on the filter paper [1] is called the stationary phase, and the solvent is called the mobile phase. Paper chromatography is a partition process between the two phases, the paper being the solid support for the aqueous stationary phase. Paper chromatography can be used for the separation of mixture. The separation can be due to the different distribution coefficient of the components in stationary phase and mobile phase.

Chromatography was primarily applied to separate colored substances, and now it may also be applied to separate colorless compounds, such as amino acids. The compound ninhydrin can react with all amino acids to produce purple products. So after developing the chromatogram, ninhydrin solution should be sprayed so that the spots, corresponding to the different amino acids, could be showed, you will then determine a R_f value for each compound.

3.36.3 Apparatus and reagents

(1) Apparatus

Chromatography jar; filter paper (5cm×20cm), capillary pipette (diameter 1mm), oven, spray box; ruler, pencil.

(2) Regents

Glutamic acid (0.015mol/L ethanol solution), leucine (0.015mol/L ethanol solution), the mixture of glutamic acid and leucine, ninhydrin (0.056mol/L acetone solution), developer: 1-butanol-acetic acid-water in the ratio of 4∶1∶5 ($V/V/V$).

3.36.4 Procedure

① Choose a precut (5cm×20cm) sheet (making sure to touch it only along the top edge) and place it on a clean sheet of notebook paper. With the short (5cm) way alined to the left and right, using a straightedge, draw a light pencil line (not ink!) from left to right. Parallel to and up from the bottom edge by 1.5cm, place 3 small pencil marks at 1.5cm intervals along this line. Label the three marks with A, A+B and B.

② Using a spare sheet of ordinary filter paper, you will practice using a capillary tube to make a spot between 0.2 and 0.3cm in diameter. Then clip the capillary tubes into the glutamic acid solution and make the spot on the sheet, applying it at the position marked with A, leucine at. the spot B and the mixture of the two at the spot A+B. Allow the paper to dry for a few minutes. If the spot is not big enough then make second application at the same positions as the first. Allow the paper to dry.

③ Pour the developer, into the chromatography jar with the depth about 1.5cm. Allow it to stand for few minutes before the separation, in order to form the saturated vapor in jar. Hang the strip in the jar with the marked end in the lower, and dip into the developer (insuring the spot do not dip into the developer and the edges of the paper not be allowed to touch the inside wall of the chromatography jar). Cover the jar tightly[2].

④ Allow the chromatogram to develop for at least an hour. When development is finished, you should remove the chromatogram and immediately mark the location of the solvent before it has a chance to dry.

⑤ Dry it and spray it with ninhydrin solution[3]. Then put your paper in an oven held at 105℃, and leave it there for 10 minutes.

⑥ Remove the chromatogram, and take it back to your desk. Circle each spot and measure the distance from the origin to the center of each spot as well as the distance from the origin to the solvent front. Calculate the R_f values, and report them to your instructor.

Notes

[1] Filter paper is nearly pure cellulose, a carbohydrate. The surface of the paper is normally covered with water molecules, which is attracted there by the many —OH (hydroxyl) groups of the cellulose molecules, both water and the hydroxyl groups are polarity and can form hydrogen bonds.

[2] When the chromatogram is being developed, the jar shouldn't be moved anyway.

[3] Avoid getting ninhydrin on your hands or clothing, it will cause stains that are difficult or impossible to remove.

3.36.5 Questions

① The baseline is the line upon which the spots of amine acids are applied. Why is it important that the depth of solvent in the bottom of the large jar not be greater than the distance from the bottom of the filter paper to the baseline?

② Why does one use pencil but not ink in marking the paper?

③ You are cautioned to avoid the edges of the paper touching the jar wall. Why?

3.37 Paper Electrophoresis of Amino Acid

3.37.1 Purpose

① To learn the principle of paper electrophoresis (PE).

② To practise the separation and identification of amino acid by PE.

3.37.2 Principle

An ion, which has position or negative charges, will move towards the electrode which has opposite charges. The quality is called electrophoresis. When using filter paper as sup-

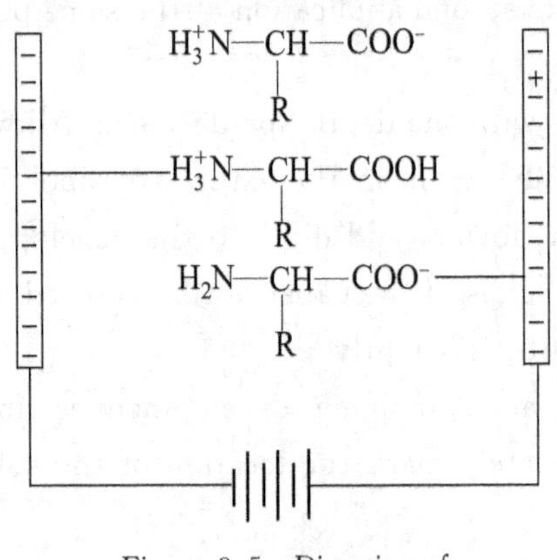

Figure 3.5 Direction of removal amino acid

porter in electrophoresis is paper electrophoresis (PE)

As we know, the exact structure of an amino acid depends on the pH. The pH at which an amino acid exists as zwitterions is known as the isoelectric point (pI). At the pI, amino acids have a zero net charge and are electrically neutral and without removal in the electric field. When the pH of solution is great than the pI of the compound, the amino acid will carry a net charge (anionic form), and will remove to the anode. Conversely, when pH of the solution is below the pI value, the amino acid exists predominantly in the cationic form (not positive charge), and to the cathode. It is shown in Figure 3.5.

According to the difference of removal direction and rate of different amino acids ions, we can separate and identify the amino acid.

3.37.3 Apparatus and Reagents

(1) Apparatus

DY-2 electrophoresis apparatus, filter paper (5cm×30cm), capillary tubes (diameter 1mm), spray box, oven, tongs, pencil.

(2) Reagents

Glutamic acid (0.015mol/L ethanol solution), arginine (0.015mol/L ethanol solution), the mixture of glutamic acid and arginine, ninhydrin (0.056mol/L acetone solution), barbital buffer solution (pH=8.9).

3.37.4 Procedure

① You will need at least three capillary tubes for use in placing the amino acid spots on the filter paper. To get best results, these tubes should be very thin, so thin that you couldn't slide a common pin inside.

② A precut (5cm×30cm) sheet of filter paper is placed on a clean sheet of notebook paper. You need to draw a thin pencil (not ink!) line in the middle of the filter paper. Place three small pencil marks and label the marks whit A, B and C at the same intervals (1.5cm) along this line, being careful not to touch the paper with your fingers.

③ Place a capillary tube in one of the sample solution, Using a spare sheet of ordinary, filter paper, practice using a capillary tube to make a spot between 0.2cm and 0.3cm in diameter; check to see just how much solution in the capillary gives the right-sized spot when the capillary tube is touched to the paper. Then dip the capillary tube into the amino acid solutions and make the spots on the large sheet, applying them at the positions marked earlier. Allow the paper to dry, for a few minutes, and then make a second application at the same position as the first. Allow the paper to dry for 5 minutes. Spot A is glutamic acid, B is arginine and C is the mixture of two formers.

④ Place the paper on the shelf of electrophoresis apparatus, dipping the both ends of the paper into the buffer. Then humidify the paper out of the pencil line about lcm with buffer until the solution permeate the line and then cover the cell.

⑤ Turn on the power, adjust voltage between 220V and 280V. About 30 minutes later, turn off the power.

⑥ When electrophoresis is finished, remove the strips out of the electrophoresis tank with a pair of tongs, and hang it in an oven at 100℃ to dry them. Then spray them with ninhydrin solution, put them in oven and keep it under 105℃ until the spots emerge, remove the strips and circle each spot.

3. 37. 5 Question

① In the experiment, the spot of glutamic acid moves towards which electrode anode or cathode? Why?

② What is the difference in the principle between paper chromatography and paper electrophoresis?

③ pI value of proline is 6. 4 and pI value of asparagine is 5. 4, when charge will be take respectively when putting them into the buffer solution with pH of 8. 9?

3. 38 Separation of Green Leaf Pigments by TLC

3. 38. 1 Purpose

① Learn separation of green leaf pigments by TLC.

② Master the operation technique of TLC.

3. 38. 2 Principle

Thin-layer chromatography (TLC) is one of the most widely used analytical techniques, at the same time it can be used for separating compounds at a scale between mg to g in the laboratory. TLC is a simple, inexpensive, fast, sensitive, and efficient method for determining the number of components in a mixture, for possibly establishing whether or not two compounds are identical, and for following the course of reaction.

Thin-layer chromatography involves the same principles as column chromatography, and it also is a form of solid-liquid adsorption chromatography. In this case, however, the solid adsorbent is spread as a thin layer (approximately 250μm) on a plate of glass or rigid plastic. A drop of the solution to be separated is placed near one edge of the plate, and the plate is placed in a container, called a developing chamber, with enough of the eluting solvent to come to a level just below the "spot" . The solvent migrates up the plate, carrying with it the components of the mixture at different rates. The result may then be seen as a series of spots on the plate, falling on a line perpendicular to the solvent level in the container. The retention factor (R_f) of a component can then be measured as indicated in thin-layer.

3. 28. 3 Apparatus and Reagents

(1) Apparatus

Mortar, separatory funnel, developing chamber.

(2) Reagents

Chromatogram sheet, petroleum ether, diethyl ether, green leaf pigments, anhydrous sodium sulfate, chloroform, ethanol.

3.38.4 Procedure

Place in mortar several fresh spinach leaves and a few milliliters of a 2 : 1 mixture of petroleum ether (b. p. 30～60℃) and ethanol, and grind the leaves well with a pestle. By means of a pipet, transfer the liquid extract to a small separatory funnel and swirl with an equal volume of water; shaking may cause formation of an emulsion. Separate and discard the lower aqueous. Repeat the water washing twice, discarding the aqueous phase each time. The water washing serves to remove the ethanol as well as other water-soluble materials that have been extracted from the leaves. Transfer the petroleum ether layer to a small flask and add 2g of anhydrous sodium sulfate. After a few minutes decant the solution from the sodium sulfate, and if the solution is not deeply and darkly colored, concentrate it by evaporating part of the petroleum ether, using a gentle stream of air.

Take a 10cm×2cm strip of silica gel chromatogram sheet. Place a spot of the pigment solution on the sheet about 1.5cm from one end, using a capillary tube to apply the spot. Avoid allowing the spot to diffuse to a diameter of no more than 2mm during application of the sample. Allow the spot to dry, and develop the chromatogram according to the general directions in the second paragraph of part 1, but use chloroform as the developing solvent.

It is sometimes possible to observe as many as eight colored spots. In order of decreasing R_f values, these spots have been identified as the carotenes (two spots, orange), chlorophyll a (blue-green), the xanthophylls (four spots, yellow), and chlorophyll b (green).

Calculate the R_f values of any spots observed on your developed plate, Also, as an aid in maintenance of a permanent record of the plate, draw to scale a picture of the developed plate in your notebook.

Notes

Have no flames in the vicinity of petroleum ether when you use it to extract the green pigments, as it is extremely flammable.

3.38.5 Questions

① In a TLC experiment why must the spot not be immersed in the solvent in the developing chamber?

② Explain why the solvent must not be allowed to evaporate from the plate during the development.

Chapter 4 Appendix

4.1 Experiments from the Literature

4.1.1 An Efficient Microscale Procedure for the Synthesis of Aspirin

To 138mg (0.001mol) of salicylic acid in a dry reaction tube is added 0.1mL of pyridine (just sufficient to dissolve it) while the tube rests in a cold water bath. One-tenth milliliter of acetyl chloride (slight excess over 0.001mol) is added in one portion. The mixture becomes viscous at this stage. The reaction tube is allowed to stand in the cold water bath for 15 min. Then 5mL of cold water is added and the mixture is shaken vigorously. It turns cloudy. Shaking is continued until white product begins to appear and at once separates. Crystals are filtered, washed with cold water, and air dried. Recrystallization from aqueous ethanol (50%) gives a yield of 33%, m.p. 133～135℃.

(Abstracted from: Journal of Chemical Education. 1998, Vol. 75, 770)

4.1.2 A Solvent-Free Claisen Condensation Reaction

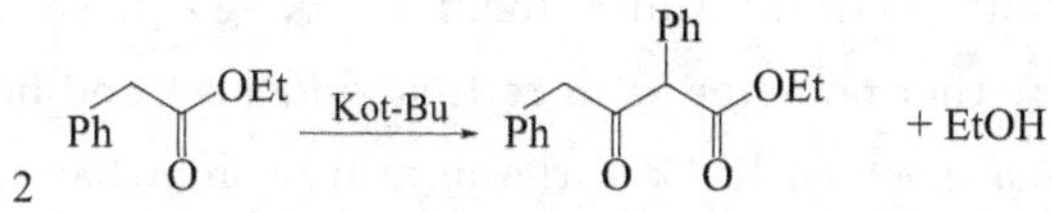

A mixture of 1.57g (14mol) of Kot-Bu, 3.28g (20mol) of ethyl phenylacetate were added to a 25mL round-bottom flask. The resulting mixture was stirred vigorously with a spatula until the contents appeared homogeneous. A reflux condenser was attached and the flask placed in a preheated hot water or steam bath at ca. 100℃. After 0.5h, the flask was removed from the heat source. The mixture was cooled to room temperature and neutralized by the slow addition of ca. 15mL of 1mol/L HCl. The residue was extracted with 2×15mL portions of ether. The organic extracts were combined, dried with $MgSO_4$, filtered, and the solvent removed by rotary evaporation. Trituration of the oily residue with 7mL of cold pentane produced a white solid. Recrystallization of the product from hot hexane afforded 2.25g of 2,4-diphenyl acetoacetate (80%), m.p. 75～78℃. Purity of the product was assessed through melting point determinations. Typical student yields are in the 50%～80% range.

(Abstracted From: Journal of Chemical Education. 2003, Vol. 80, 1446)

4.1.3 Convenient Synthesis of a Lactone, γ-Butyrolactone

Place 1.5g (11.9mmol) of γ-hydroxybutyric acid sodium salt into a 5mL conical vial that contains a spin vane. Carefully add 1.5mL of 9mol/L H_2SO_4, attach an air condenser, and reflux the mixture for 15 minutes. Upon cooling, crystals of sodium sulfate may form.

Extract the reaction mixture with 1.5mL of CH_2Cl_2. Shake well with frequent venting. Allow the phases to separate and transfer the upper layer, using a filter-tip pipet, to a glass centrifuge tube. Repeat the extraction with a second 1.5mL portion of CH_2Cl_2 and transfer the upper layer to the centrifuge tube as before.

Dry the organic phase in the corked centrifuge tube with three microscoopula tips (about 0.25g) of anhydrous Na_2SO_4 (at least 15 minutes).

Transfer the dried CH_2Cl_2/lactone extract, using a clean filter-tip pipet, to a preweighed vial. Evaporate the CH_2Cl_2 using a hot-plate and a gentle stream of N_2 until the weight of the vial and liquid remains constant, then reweigh.

Determine the percent yield in the usual manner, obtain an IR spectrum, and compare it to an authentic IR spectrum of γ-butyrolactone.

(Abstracted from: Journal of Chemical Education. 1998, Vol. 75, 84)

4.1.4 Polyester (PET) Synthesis

H_3CO–CO–C_6H_4–CO–OCH_3 + 2 HO–CH_2CH_2–OH ⇌ HO–CH_2CH_2–O–CO–C_6H_4–CO–O–CH_2CH_2–OH + $2CH_3OH$

nHO–CH_2CH_2–O–CO–C_6H_4–CO–O–CH_2CH_2–OH $\xrightarrow[\text{heat}]{H^+}$ $\left(\text{CO–}C_6H_4\text{–CO–O–}CH_2CH_2\text{–O} \right)_n$ + n HO–CH_2CH_2–OH

Place dimethyl terephthalate (5.0g), ethane diol (80mL) and some antibumping granules in a 250mL round-bottom flask. Add a small (≈0.1g) piece of sodium metal (CAUTION!). Fit a Y adapter, thermometer, and reflux condenser and heat the mixture to reflux for 45 minutes using a heating mantle. Cool the mixture somewhat and modify the apparatus for distillation, leaving the Y adapter in place for fractional distillation. Distill off the methanol, stopping the collection when the head temperature reaches 180℃. Decant the solution from antibumping granules and allow it to cool. Collect the crystals that form by suction filtration. Dry a sample thoroughly between filter papers and place it in a crucible or on a microscope slide. Heat this strongly on a hot-plate, carefully removing samples from time to time to check their physical properties as they cool. Repeat with another dry sample, this time adding a small crystal of p-toluene sulfonic acid to the molten material. Wash the bulk sample with water and dry thoroughly before submitting a sample for ^{1}H NMR analysis.

(Abstracted from: Journal of Chemical Education. 1999, Vol. 76, 236～237)

4.1.5 A Safe Simple Halogenation Experiment

CH_3–CH(CH_3)–CH(CH_3)–CH_3 $\xrightarrow[\text{light}]{Br_2}$ CH_3–CBr(CH_3)–CH(CH_3)–CH_3 + CH_3–CBr(CH_3)–CBr(CH_3)–CH_3

WARNING: Bromine causes severe burns and should be dispensed by the instructor wearing chemical-resistant gloves. A 10% solution of sodium thiosulfate should be kept nearby in case of a spill or skin Contact. Only after the reaction is complete (bromine color gone)

are students allowed to contact the reaction vessel. Care must still be given to the remaining HBr fumes and the alkyl halides produced in the reaction.

Into a wide 50mL test tube or centrifuge tube is placed 10mL of 2,3-dimethylbutane followed by 1.0mL of bromine. Care is needed in handling bromine; we have the instructor dispense the bromine from a tilting dispenser. The reaction mixture is swirled briefly and the test tube is propped up in a 400mL beaker pointed toward the light source. After a few minutes, copious mounts of "HBr" can be seen and the bromine color eventually fades to a light yellow. When the reaction is complete, the test tube should be loosely corked to diminish water condensation as ice is added to the surrounding beaker. Cooling the solution to 0℃ will precipitate a white solid, which can be collected by suction filtration. This should be done in the hood, since there will still be some HBr in the solution. The product should be air-dried for a few minutes on the filter paper before taking its melting point and its NMR spectrum. The reported melting point of 2,3-dibromo-2,3-dimethylbutane is 169℃ (scaled tube). Student yields are usually more than 50% and melting points are within a few degrees of the literature value when taken in sealed tubes.

(Abstracted from: Journal of Chemical Education. 1999, Vol. 76, 534)

4.1.6 Microwave Microscale Experiment—2-Naphthyl Acetate from 2-Naphthol

OH $\xrightarrow{(CH_3CO)_2O}$ $OCOCH_3$

Experimental Procedure

2-Naphthol (288mg, 2mmol) and 1mL of acetic anhydride are mixed in a 50mL beaker. The beaker is covered with a watch glass and irradiated in the microwave for 4min at 50% power. The reaction mixture is poured into 20mL of water. After the mixture has been stirred to hydrolyze the excess acetic anhydride, the solid is collected and recrystallized in 50% ethanol to give a 60%~80% yield of product (m.p. 67.0~67.5℃).

(Abstracted from: Journal of Chemical Education. 1996, Vol. 73, A105)

4.2 List of the Elements with Their Symbols and Atomic Masses*

Element	Symbol	Atomic Number	Atomic Mass†
Actinium	Ac	89	(227)
Aluminum	Al	13	26.98
Americium	Am	95	(243)
Antimony	Sb	51	121.8
Argon	Ar	18	39.95
Arsenic	As	33	74.92
Astatine	At	85	(210)
Barium	Ba	56	137.3
Berkelium	Bk	97	(247)
Beryllium	Be	4	9.012
Bismuth	Bi	83	209.0

续表

Element	Symbol	Atomic Number	Atomic Mass†
Boron	B	5	10.81
Bromine	Br	35	79.90
Cadmium	Cd	48	112.4
Calcium	Ca	20	40.08
Californium	Cf	98	(249)
Carbon	C	6	12.01
Cerium	Ce	58	140.1
Cesium	Cs	55	132.9
Chlorine	Cl	17	35.45
Chromium	Cr	24	52.00
Cobalt	Co	27	58.93
Copper	Cu	29	63.55
Curium	Cm	96	(247)
Dysprosium	Dy	66	162.5
Einsteinium	Es	99	(254)
Erbium	Er	68	167.3
Europium	Eu	63	152.0
Fermium	Fm	100	(253)
Fluorine	F	9	19.00
Francium	Fr	87	(223)
Gadolinium	Gd	64	157.3
Gallium	Ga	31	69.72
Germanium	Ge	32	72.59
Gold	Au	79	197.0
Hafnium	Hf	72	178.5
Hahnium	Ha	105	(260)
Hassium	Hs	108	(265)
Helium	He	2	4.003
Holmium	Ho	67	164.9
Hydrogen	H	1	1.008
Indium	In	49	114.8
Iodine	I	53	126.9
Iridium	Ir	77	192.2
Iron	Fe	26	55.85
Krypton	Kr	36	83.80
Lanthanum	La	57	138.9
Lawrencium	Lr	103	(257)
Lead	Pb	82	207.2
Lithium	Li	3	6.941
Lutetium	Lu	71	175.0
Magnesium	Mg	12	24.31
Manganese	Mn	25	54.94
Meitnerium	Mt	109	(266)
Mendelevium	Md	101	(256)
Mercury	Hg	80	200.6
Molybdenum	Mo	42	95.94
Neodymium	Nd	60	144.2
Neon	Ne	10	20.18
Neptumium	Np	93	(237)
Nickel	Ni	28	58.69
Nielsbohrium	Ns	107	(262)
Niobium	Nb	41	92.91
Nitrogen	N	7	14.01
Nobelium	No	102	(253)

续表

Element	Symbol	Atomic Number	Atomic Mass†
Osmium	Os	76	190.2
Oxygen	O	8	16.00
Palladium	Pd	46	106.4
Phosphorus	P	15	30.97
Platinum	Pt	78	195.1
Plutonium	Pu	94	(242)
Polonium	Po	84	(210)
Potassium	K	19	39.10
Praseodymium	Pr	59	140.9
Promethium	Pm	61	(147)
Protactinium	Pa	91	(231)
Radium	Ra	88	(226)
Radon	Rn	86	(222)
Rhenium	Re	75	186.2
Rhodium	Rh	45	102.9
Rubidium	Rb	37	85.47
Ruthenium	Ru	44	101.1
Rutherfordium	Rf	104	(257)
Samarium	Sm	62	150.4
Scandium	Sc	21	44.96
Seaborgium	Sg	106	(263)
Selenium	Se	34	78.96
Silicon	Si	14	28.09
Silver	Ag	47	107.9
Sodium	Na	11	22.99
Strontium	Sr	38	87.62
Sulfur	S	16	32.07
Tantalum	Ta	73	180.9
Technetium	Tc	43	(99)
Tellurium	Te	52	127.6
Terbium	Tb	65	158.9
Thallium	Tl	81	204.4
Thorium	Th	90	232.0
Thulium	Tm	69	168.9
Tin	Sn	50	118.7
Titanium	Ti	22	47.88
Tungsten	W	74	183.9
Uranium	U	92	238.0
Ununbium	Uub	112	(277)
Ununbexium	Uuh	116	(289)
Ununnilium	Uun	110	(269)
Ununoctium	Uuo	118	(293)
Ununquadium	Uuq	114	(285)
Unununium	Uuu	111	(272)
Vanadium	V	23	50.94
Xenon	Xe	54	131.3
Ytterbium	Yb	70	173.0
Yttrium	Y	39	88.91
Zinc	Zn	30	65.39
Zirconium	Zr	40	91.22

* All atomic masses have four significant figures. These values are recommended by the Committee on Teaching of Chemistry, International Union of Pure and Applied Chemistry.

† Approximate values of atomic masses for radioactive elements are given in parentheses.

Katherine J, Derniston et al. General. Organic, and Biochemistry 3rd ed. Published by Mc. Graw-Hill Companies, Inc., New York. NY10020, 1997. (www.mhhe.com)

4.3 Main Families of Organic Compounds

Families	Specific Example	IUPAC Name	Common Name	General Formula	Functional Group
Alkane	CH_3CH_3	Ethane	Ethane	RH	C—C
Alkene	$H_2C=CH_2$	Ethene or Ethylene	Ethylene	$R_2C=CR_2$ (R=H,R′)	C=C
Alkyne	$HC\equiv CH$	Ethyne or Acetylene	Acetylene	$RC\equiv CR$ (R=H,R′)	$-C\equiv C-$
Arene		Benzene	Benzene	ArH	Aromatic Ring
Haloalkane	CH_3CH_2Cl	Chloro Ethane	Ethyl chloride	RX	—X
Alcohol	CH_3CH_2OH	Ethanol	Alcohol	ROH	—OH
Ether	CH_3OCH_3	Methoxymethane	Dimethyl ether	ROR	C—O—C
Amine	CH_3NH_2	Methanamine	Methylamine	RNH_2 R_2NH R_3N	$C-NR_2$
Aldehyde	CH_3CHO	Ethanal	Acetal-dehyde	RCHO	$-C(=O)H$
Ketone	CH_3COCH_3	Propanone	Acetone	RCOR	$-C(=O)-$
Carboxylic Acid	CH_3COOH	Ethanoic Acid	Acetic acid	RCOOH	$-C(=O)OH$
Ester	CH_3COOCH_3	Methyl Ethanoate	Methyl acetate	RCOOR′	$-C(=O)OR$
Amide	CH_3CONH_2	Ethanamide	Acetamide	$RCONR'_2$ (R′=H,R″)	$-C(=O)NR'_2$

4.4 Boling Point and Density of Some Common Organic Reagents

Name	Boling Point/℃	Density(d_4^{20})
Methanol	64.96	0.7914
Ethanol	78.5	0.7893
Diethyl ether	34.51	0.71378
Acetone	56.2	0.7899
Acetic acid	117.9	1.0492
Acetic anhydride	139.55	1.0820
Ethyl Acetate	77.06	0.9003
Benzene	80.1	0.87865
Toluene	110.6	0.8669
Dimethyl benzene	140	
Chloroform	61.7	1.4832
Carbon tetrachloride	76.54	1.5940
Nitrobenzene	210.8	1.2037
n-Butanol	117.25	0.8098

第1章　有机化学实验基础知识

有机化学实验是一门以实验为基础的学科，随着新的实验技术不断出现，这门实验课程正在向用量少、效率高、绿色化的方向发展。通过有机化学实验课程的学习，使学生能够掌握有机化学实验的基本原理，有机物的合成、分离、鉴定的一般方法；加深对有机化学理论知识的理解，培养学生“预习（包括查阅文献）—准备—实验—记录—总结”的实验习惯，以及严谨的科学态度和工作作风。

1.1　有机化学实验的基本规则

为确保有机化学实验有条不紊、安全地进行，必须遵循以下规则。

① 熟悉实验室安全守则，学会正确使用水、电、煤气、通风橱、灭火器等，了解实验事故的一般处理方法。做好实验预习工作，了解所用药品的危害性及安全操作方法，按操作规程，小心使用有关实验仪器和设备。

② 实验前，应认真清点、检查玻璃仪器；实验中，安全合理地使用玻璃仪器；实验后，洗净并妥善保管玻璃仪器，尤其应学会玻璃仪器的洗涤方法。

③ 实验时，要保持实验室和桌面的清洁，认真操作，遵守实验纪律，严格按照实验中所规定的实验步骤、试剂规格及用量来进行。若要改变，需经指导教师同意。

④ 实验药品使用前，应仔细阅读药品标签，按需取用，避免浪费；取完药品后要迅速盖上瓶塞，避免盖错瓶塞、污染药品。不要任意更换实验室常用仪器（如天平、干燥器、折光仪等）和常用药品的摆放位置。

⑤ 整个实验操作过程中要集中思想，避免大声喧哗，不要在实验室吃东西。

⑥ 实验中和实验后，各类固体废物和液体废物应分别放入指定的废物收集器中。

⑦ 离开实验室前，应检查水、电、煤气是否安全关闭。

1.2　常用的玻璃仪器和实验装置

1.2.1　玻璃仪器

参见英文部分。

1.2.2　实验装置

参见英文部分。

1.2.3　玻璃仪器的使用及注意事项

① 使用时要轻拿轻放，以免弄碎。

② 除烧杯、烧瓶和试管外，均不能用火直接加热。

③ 锥形瓶、平底烧瓶不耐压，不能用于减压系统。

④ 带活塞的玻璃器皿用过洗净后，要在活塞与磨口之间垫上纸片。

⑤ 温度计的水银球玻璃很薄，易碎，使用时应小心。不能将温度计当搅拌棒使用；温度计使用后应先冷却再冲洗，以免破裂；测量范围不得超出温度计刻度范围。

⑥ 温度计若不慎破碎，应立即将撒落的水银用硫黄覆盖，碎温度计应插入盛有硫黄的容器中。

1.2.4 实验装置的安装及注意事项

① 所用玻璃仪器和配件要干净，大小要合适。

② 安装实验装置时应按照从下向上、从左到右原则，逐个装配。

③ 拆卸时，则按从右到左、从上到下原则，逐个拆除。

④ 常压下进行的反应装置，应与大气相通，不能密封。

⑤ 实验装置要求做到严密、正确、整齐、稳妥，磨口连接处要呈一直线。

1.2.5 玻璃仪器的清洗

玻璃仪器用毕后应立即清洗，一般的清洗方法是将玻璃仪器和毛刷淋湿，蘸取肥皂粉或洗涤剂，洗刷玻璃器皿的内外壁，除去污物后用水冲洗；当洁净度要求较高时，可依次用洗涤剂、蒸馏水（或去离子水）清洗；也可用超声波振荡仪来清洗。

坚决反对盲目使用各种化学试剂或有机溶剂来清洗玻璃器皿，这样不仅造成浪费，而且可能带来危险，对环境产生污染。

1.2.6 玻璃仪器的干燥

干燥玻璃仪器的方法通常有以下几种。

（1）自然干燥　将仪器倒置，使水自然流下，晾干。

（2）烘干　将仪器放入烘箱内烘干，仪器口朝上；也可用气流干燥器烘干或用电吹风吹干。

（3）使用有机溶剂助干燥　急用时可用有机溶剂助干，用少量95%乙醇或丙酮荡涤，把溶剂倒回至回收瓶中，然后用电吹风吹干。

附：简单玻璃工操作

玻璃管的切割、弯制及毛细管的制作等都是有机化学实验中最基本的操作。

（1）切割玻璃管和玻璃棒

① 用小的砂轮片在所需截断的地方垂直于玻璃管（棒），向一个方向锉一个稍深的凹痕（注意不要来回锉，另外玻璃管应是干燥的）。

② 两手握住玻璃管（棒），用大拇指顶住锉痕背面的两边，轻轻向前推，同时双手向两边拉，玻璃管（棒）即可平整断开［注意：折断玻璃管（棒）时，应远离眼睛］。

③ 断口处必须淬火，圆口，即把断口处放在煤气灯的火焰上来回旋转几下，以除去断口面的快口[1]。

（2）弯玻璃管

① 将干燥的玻璃管倾斜一定角度，放在煤气灯的强火焰上灼烧，双手均匀地向一个方向转动。

② 当加热部位呈红色开始软化时，即从火焰中取出，轻轻弯成所需的角度。

③ 加热或弯曲管子时不能扭动，以避免弯管不在同一个平面上的现象。

④ 弯管时要避免弯角内侧出现瘪陷，弯好后应检查角度是否正确，整个弯管是否在同一平面上[2]，然后放在石棉网上冷却（不要直接放在桌面上）。

（3）实验

① 取数根废旧的玻璃管练习切割、加热、转动、弯制等操作，掌握操作要点。

② 搅棒制作：取一根直径为5mm、长度约20cm的玻璃棒，两端分别在火焰上烧圆，作为搅棒。

③ 玻璃管的弯制：取两根直径为7mm的玻璃管，弯成90°、70°角的玻璃管各一支[3]。

注释：

[1] 有时断口面不平整，易割伤手，要小心。

[2] 玻璃管弯好后，亦可在小火上进行退火，然后再放于石棉网上。

[3] 玻璃管的长度视用途而定。

1.3　有机化学实验的安全知识

有机化学实验多采用玻璃仪器、实验试剂和电器设备等，如果操作不当会对人体、环境造成伤害，实验试剂往往具有易燃、易爆、易挥发、易腐蚀、毒性高等特点，玻璃仪器与电器设备使用不当亦可发生意外事故。因此，有机化学实验室是一个潜在的高危险性的场所。

1.3.1　防火

实验操作要规范，实验装置要正确，对易燃、易爆、易挥发的实验药品要远离明火，不可随意丢弃，实验后应专门回收。若一旦发生火灾，应先切断电源、煤气，移去易燃、易爆试剂，再采取适当方法灭火，如：灭火器，石棉网或沙土覆盖，或用水冲等。

1.3.2　防爆

仪器装置要正确，常压蒸馏及回流时，整个系统不能密闭；减压蒸馏时，应事先检查玻璃仪器是否能承受系统的压力；若在加热后发现未放沸石，应停止加热，冷却后再补加；冷凝水要保持畅通。

有些有机物遇氧化剂会发生猛烈的爆炸或燃烧，操作或存放应格外小心。

1.3.3　防中毒

绝大多数有机化学试剂都有不同程度的毒性，对有刺激性或者产生有毒气体的实验，应尽量安排在通风橱或有排风系统的环境中进行，或采用气体吸收装置。有毒或有较强腐蚀性的药品应严格按照有关操作规程进行，不能用手直接拿或接触这类化学药品，不得入口或接触伤口，亦不可随便倒入下水道。

实验中若发现有头晕、头痛等中毒症状，应立即转移到空气新鲜的地方休息，严重者应立即送往医院。

1.3.4　防化学灼伤

强酸、强碱和溴等化学药品接触皮肤均可引起灼伤，使用时应格外小心。一旦发生这类情况应立即用大量水冲洗，再用如下方法处理。

酸灼伤：眼睛灼伤用1%$NaHCO_3$溶液清洗；皮肤灼伤用5%$NaHCO_3$溶液清洗。

碱灼伤：眼睛灼伤用1%硼酸溶液清洗；皮肤灼伤用1%～2%醋酸溶液清洗。

溴灼伤：立即用酒精洗涤，再涂上甘油，或敷上烫伤油膏。

灼伤较严重者经急救后速去医院治疗。

1.3.5 防割伤和烫伤

在玻璃仪器的使用和玻璃工的操作中，常因操作或使用不当而发生割伤和烫伤现象。若发生此类现象，可用如下方法处理。

割伤：先要取出玻璃片，用蒸馏水或双氧水清洗伤口，涂上红药水，再用纱布包扎；若伤口严重，应在伤口上方用纱布扎紧，急送医院。

烫伤：轻者涂烫伤膏，重者涂烫伤膏后立即送医院。

1.4 有机化学实验废物的处置

在有机化学实验中和实验结束后往往会产生各种固体、液体等废物，为遵守国家的环保法规，避免或减少对环境的危害，可采用如下处理方法。

① 所有实验废物应按固体、液体或有害、无害等分类收集于不同的容器中，对一些难处理的有害废物可送环保部门专门处理。

② 少量的酸（如盐酸、硫酸、硝酸等）或碱（如氢氧化钠、氢氧化钾等）在倒入下水道之前必须被中和，并用水稀释。

③ 有机溶剂必须倒入带有标签的废物回收容器中，并存放在通风处。

④ 对无害的固体废物，如：滤纸、碎玻璃、软木塞、氧化铝、硅胶、硫酸镁、氯化钙，可直接倒入普通的废物箱中，不应与其他有害固体废物相混；对有害固体废物应放入带有标签的广口瓶中。

⑤ 对能与水发生剧烈反应的化学品，处置之前要用适当的方法在通风橱内进行分解。

⑥ 对可能致癌的物质，处理起来应格外小心。避免与手接触。

1.5 常用有机溶剂及纯化

（1）乙醚（$C_2H_5OC_2H_5$）

相对分子质量 74.1，沸点（b.p.）34.5℃，相对密度（d_4^{20}）0.71。乙醚沸点低易挥发、易燃，使用乙醚时严禁明火。乙醚几乎能和所有的有机溶剂任意混合，在水中的溶解度约10%。乙醚久置易产生过氧化物，蒸馏久置的乙醚时切忌蒸干，以免因过氧化物引起爆炸。乙醚应贮存于密闭容器中并放阴凉处。

① 过氧化物的检验　取少量乙醚，加等体积的2%碘化钾水溶液和几滴稀硫酸，振摇，再加1滴淀粉试液，呈紫蓝色即表示有过氧化物存在。

② 过氧化物的除去　用酸性硫酸亚铁溶液（110mL 水，6mL 浓硫酸，60g 硫酸亚铁）洗涤乙醚可除去过氧化物。然后用水洗涤，用无水氯化钙干燥，蒸馏得纯乙醚。

③ 无水乙醚的制备　将 100mL 乙醚放在干燥锥形瓶中，加入 20～25g 无水氯化钙，加塞放置 1d 以上，并间断摇动，然后蒸馏收集 33～37℃馏分。用压钠机将 1g 金属钠直接压成钠丝放入盛乙醚的瓶中，用带有氯化钙干燥管的木塞塞住，或在木塞中插一末端拉成毛细管的玻璃管。这样既可防止潮气浸入，又可使产生的气体逸出。放置至无气泡发生即可使用。若钠丝表面已变黄、变粗，须再蒸一次，然后再压入钠丝。

(2) 乙醇 (C_2H_5OH)

相对分子质量 46.1，沸点 78.5℃，相对密度 (d_4^{20})0.789。乙醇为具有酒味的无色透明液体，易燃，能与水任意混合，对人体的毒性较低，许多极性和极性较小的有机化合物能溶解在乙醇中，因此乙醇是重结晶有机化合物的良好溶剂。乙醇能与水形成共沸物(b.p.78.2℃)，用一般的分馏法不能完全除去其中的水。市售乙醇的含量为95%。

① 无水乙醇的制备（含量为 99.5%）　在 250mL 圆底烧瓶中，放入 45g 生石灰、100mL (95%) 乙醇，装上回流冷凝器（上接一无水氯化钙干燥管），在水浴上回流 2～3h，然后改为蒸馏装置蒸馏，弃去少量前馏分后收集得无水乙醇。

② 绝对乙醇的制备（含量为 99.95%）

a. 用金属钠制备。在 250mL 圆底烧瓶中，将 2g 金属钠加入 100mL 纯度至少是 99%的乙醇中，加几粒沸石，装上球形冷凝器（上接一个无水氯化钙干燥管），回流 30min。再改成蒸馏装置蒸馏，收集得绝对乙醇。若要制备纯度更高的绝对乙醇，则可在回流 30min 后，加入 4g 邻苯二甲酸二乙酯，再回流 10min，然后改成蒸馏装置蒸馏，收集产品即得。

b. 用金属镁制备。装置同上，在 250mL 圆底烧瓶中加入 0.6g 干燥镁条（或镁屑）和 10mL 99.5%乙醇。在水浴上微热后移去热源，立即投入几小粒碘粒加速反应进行（注意不要摇动）。不久碘粒周围即发生反应（如反应太慢可加热或补加碘粒），慢慢扩大，最后可达到相当激烈的程度。当全部镁条反应完毕后，加入 100mL 99.5%乙醇和几粒沸石，回流 1h，以下操作同 a。

(3) 丙酮 (CH_3COCH_3)

相对分子质量 58.1，沸点 56.5℃，相对密度 (d_4^{20})0.789。丙酮易燃，溶于水，与多种有机溶剂能任意混合。丙酮对有机化合物有较好的溶解度，是精制有机物质的良好溶剂。丙酮可用无水硫酸钙或无水碳酸钾干燥去水。

工业丙酮常含有醛或其他还原性杂质。加入少量高锰酸钾回流可将杂质除去。若高锰酸钾紫色很快褪去，再加入少量高锰酸钾继续回流，直至紫色不褪为止。然后将丙酮蒸出，用无水硫酸钙或无水硫酸钾干燥，过滤，蒸馏即得较纯的丙酮。

(4) 苯 (C_6H_6)

相对分子质量 78.1，沸点 80℃，相对密度 (d_4^{20})0.8787。苯是无色透明的液体，易燃，与水不混溶。苯是非极性溶剂，常用来提取、重结晶和层析有机化合物。苯和水能形成共沸混合物（b.p.69℃，含水量为 9%），故常利用苯的这种性质来除去反应中生成的水。苯蒸气有毒，长期接触会引起慢性中毒，主要表现为破坏人体造血功能。

工业苯常含有少量噻吩和水，不能用分馏方法除去。

① 噻吩的检验　取 1mL 苯加入 2mL 溶有 2mg 吲哚醌的浓硫酸溶液，振荡片刻，若酸层呈墨绿色或蓝色，即表示有噻吩存在。

② 噻吩和水的除去　将苯和 1/10 体积的浓硫酸振摇，使噻吩形成噻吩-2-磺酸，重复几次，直到检验无噻吩为止。然后分去酸层，苯层用水洗涤至中性，用无水氯化钙干燥后蒸馏，收集 80℃馏分。再压入金属钠丝即成无噻吩、无水的苯。

(5) 甲苯 ($CH_3C_6H_5$)

相对分子质量 92.1，沸点 111℃，相对密度 (d_4^{20})0.866。甲苯为易燃无色液体，它的毒性较苯小。甲苯几乎不溶于水，它也能和水形成共沸混合物（b.p.85℃）。共沸混合物中约含 20%的水，因此甲苯的去水量相当大，加上它本身又是一个较好的溶剂，因此在实验中经常用甲苯除去反应中生成的水。

普通甲苯中可能含有少量甲基噻吩。除去甲基噻吩是在 1000mL 甲苯中加入 100mL 浓硫酸，振摇约 30min（温度不要超过 30℃），除去酸层，甲苯层用水洗至中性。用无水氯化钙干燥，过滤，蒸馏，即可得纯品。

（6）甲醇（CH_3OH）

相对分子质量 32.04，沸点 64.6℃，相对密度（d_4^{20}）0.79。甲醇为无色透明的易燃性液体，它与水及许多极性溶剂任意混合，是实验中常用的良好溶剂。甲醇剧毒，饮用后引起眼盲甚至死亡。甲醇与水不形成共沸物，可直接用高效分馏法制备无水甲醇，或用镁处理后制备无水甲醇（参考无水乙醇的制备）。

（7）乙酸乙酯（$CH_3COOC_2H_5$）

相对分子质量 88.1，沸点 77.2℃，相对密度（d_4^{20}）0.90。乙酸乙酯为无色易燃液体，能与多数有机溶剂混合，100mL 水中能溶解 8.6g 乙酸乙酯。乙酸乙酯与水的共沸物沸点为 70.38℃。乙酸乙酯是许多有机化合物的良好溶剂，它能与胺类起反应，精制这些胺类化合物时不能用乙酸乙酯作溶剂。

不纯的乙酸乙酯常含少量的乙酸和乙醇，可依次用 5% 碳酸钠溶液洗、水洗。然后用无水硫酸钠或无水硫酸镁干燥，蒸馏即得较纯的乙酸乙酯。

（8）二甲亚砜（C_2H_5SO）

相对分子质量 78.1，沸点 189℃，相对密度（d_4^{20}）1.10。二甲亚砜是高极性的非质子溶剂，能与水互溶，广泛用于有机反应和光谱分析中的溶剂。二甲亚砜易吸潮，常压蒸馏时会分解。如要无水二甲亚砜，可以用活性氧化铝、氧化钡或硫酸钙干燥过夜，滤去干燥后减压收集 75～76℃/12mmHg（1mmHg=133.325Pa）馏分，放入分子筛储存待用。

（9）三氯甲烷（$CHCl_3$）

相对分子质量 119.4，沸点 61.7℃，相对密度（d_4^{20}）1.48。三氯甲烷为无色透明液体，微溶于水，蒸气不燃烧。三氯甲烷能溶解许多有机化合物，实验中可用它萃取和精制有机化合物。三氯甲烷能与水形成共沸物，沸点 61℃。

三氯甲烷不能和碱性试剂或溶液接触，因碱能使三氯甲烷分解为二氯卡宾，有时这种分解反应很剧烈。一般不用三氯甲烷作胺类的溶剂，也不能用三氯甲烷来提取强碱性物质。三氯甲烷具有毒性，大量接触后会引起肝肾损伤和心律不齐。三氯甲烷暴露在日光和空气中会慢慢氧化为剧毒的光气。一般三氯甲烷中均加入 0.5%～1% 乙醇作为稳定剂。

除去三氯甲烷中的乙醇可用其体积一半的水洗涤 5～6 次，然后用无水氯化钙干燥，再蒸馏。三氯甲烷纯品要放置于暗处。三氯甲烷不能用金属钠干燥，因为会发生爆炸。

（10）二氯甲烷（CH_2Cl_2）

相对分子质量 84.9，沸点 40℃，相对密度（d_4^{20}）1.32。二氯甲烷不易燃，与水不溶。它是许多有机化合物的优良溶剂，因沸点低，可在低温下浓缩。二氯甲烷是氯代甲烷中毒性最低的一种溶剂，如有可能可用它来代替三氯甲烷。

用水、碳酸钠溶液洗涤不纯的二氯甲烷，然后用无水氯化钙干燥，分馏后可得较纯的二氯甲烷。

（11）石油醚

石油醚是低分子量烷烃类的混合物。市售石油醚按其沸程可分为 30～60℃、60～90℃和 90～120℃三种规格。石油醚中含有少量不饱和烃，因这些不饱和烃的沸点与烷烃相近，因此用蒸馏法不能将它们分离除去。

石油醚的精制通常是用其体积1/10的浓硫酸将石油醚洗涤两三次，再用10%的硫酸加入高锰酸钾配成的饱和溶液洗涤，直至水层中的紫色不再消失为止。然后再用水洗，经无水氯化钙干燥后蒸馏。如要绝对干燥的石油醚则可压入钠丝（参考无水乙醚的制备）。

(12) 正己烷（C_6H_{14}）

相对分子质量86.2，沸点69℃，相对密度（d_4^{20}）0.66。正己烷为无色透明液体，易燃，不溶于水。常用正己烷来提取、精制和层析有机化合物。

正己烷中的主要杂质是烯烃和芳香族化合物，可按下述方法将这些杂质除去。即将正己烷与浓硫酸的混合液进行搅拌，分去浓硫酸后再用0.1mol/L高锰酸钾溶液和10%硫酸溶液进行搅拌，分出上层正己烷液层，用水洗涤至中性，用无水氯化钙干燥。干燥后的正己烷再经蒸馏，馏出液中压入金属钠丝除去微量的水分，即得到纯的无水正己烷。

(13) 环己烷（C_6H_{12}）

相对分子质量84.2，沸点80.7℃，相对密度（d_4^{20}）0.778。环己烷是无色透明液体，易燃，不溶于水，它能与多种有机溶剂混溶。环己烷中主要杂质是苯和一些不饱和烃，可用冷的浓硫酸和浓硝酸混合液洗涤数次除去。分去酸层后的环己烷用水洗涤至中性，无水氯化钙干燥后分馏，再压入金属钠丝除去微量水分成为无水环己烷。

(14) 四氢呋喃（THF，C_4H_8O）

相对分子质量72.1，沸点66℃，相对密度（d_4^{20}）0.89。四氢呋喃为可燃性无色液体，可与水或其他有机溶剂任意混合，是有机反应的良好溶剂。四氢呋喃含水，久贮后可能含有过氧化物，在加碱处理或蒸馏近干时会引起爆炸。

① 过氧化物的检验　将四氢呋喃加入到等体积的2%碘化钾溶液和淀粉溶液中，再加入几滴酸摇匀，如呈蓝或紫色，表示有过氧化物存在。

② 无水四氢呋喃的制备　用氢化锂铝在隔绝潮气下回流（通常1000mL约用2～4g氢化锂铝）除去其中的水和过氧化物。处理过的四氢呋喃中加入钠丝和二苯酮，应出现深蓝色的二苯酮钠，且加热回流蓝色不褪。在氮气保护下蒸馏，收集66℃的馏分。

在处理四氢呋喃时，应先取少量进行试验，待确定其只含有少量水分和过氧化物后，才可进行大量的处理。如果四氢呋喃含水量较多，则可将固体氢氧化钾或其浓溶液与四氢呋喃回流加热，此时四氢呋喃溶液将出现红色难溶性树脂状物质，然后蒸出四氢呋喃再作处理。

(15) 吡啶（C_5H_5N）

相对分子质量79.1，沸点115℃，密度（d_4^{20}）0.98。吡啶吸水力强，能与水、醇和醚任意混合。吡啶与粒状氢氧化钾（钠）一起回流，然后隔绝潮气蒸出即得无水吡啶。

(16) 乙腈（CH_3CN）

相对分子质量41.1，沸点82℃，相对密度（d_4^{20}）0.78。乙腈为易燃性液体，能与水混合。乙腈是许多有机物的良好溶剂。乙腈可用5%五氧化二磷一起蒸馏而精制。

(17) *N*,*N*-二甲基甲酰胺（DMF，C_3H_7NO）

相对分子质量73.1，沸点153℃，相对密度（d_4^{20}）0.95。*N*,*N*-二甲基甲酰胺是可燃性液体，能与水任意混合。它对许多有机物的溶解度均较大，但很少用于有机物质的精制。吸入过多的DMF的蒸气后会引起恶心呕吐。DMF和其蒸气易通过皮肤被吸收。

N,*N*-二甲基甲酰胺含有少量水分，在常压蒸馏时部分DMF会分解，若有酸或碱存在，分解加快。因此最好用硫酸钙、硫酸镁、氧化钡、硅胶或分子筛干燥后，减压蒸馏收集

76℃/36mmHg 馏分。如含水较多，可加入 1/10 体积的苯，在常压 80℃以下蒸去水和苯，然后用硫酸镁或氧化钡干燥，再进行减压蒸馏。

（18）二氧六环（$C_4H_8O_2$）

相对分子质量 88.1，沸点 101℃/750mmHg，相对密度（d_4^{20}）1.03。二氧六环中常含有少量乙醛、缩醛和水，常用下列方法精制：

在 500mL 二氧六环中加入 7mL 浓盐酸和 50mL 水，在通风橱中加热回流 12h，回流时缓慢地将氮气通入溶液以除去乙醛。待溶液冷却后加入粒状氢氧化钾直至不再溶解。分去水，有机层再加入粒状氢氧化钾振摇除去痕量水。将有机层放入干燥的圆底烧瓶中，加入金属钠回流 10～12h，使金属钠最终保持光亮，若不是这样，可以再加入金属钠以同样方式处理。最后蒸馏收集 101℃/750mmHg 馏分。

1.6 实验预习记录和实验报告

（1）实验预习记录　在进入实验室之前必须通过预习有关化学实验的理论知识及相关信息等来做好充分的准备。如果预习充分，实验将会省时高效。通常实验预习、记录包括如下几个部分：

①实验题目、原理、仪器和实验过程；

②实验注意事项；

③实验中存在的危险。

（2）实验报告　通常实验报告应包括下面几个部分：

①实验题目；

②实验目的；实验的主要操作或技术；

③实验原理；化学反应和反应机理等；

④实验仪器和主要试剂；

⑤实验过程；实验步骤和现象记录；

⑥结果和讨论；数据处理及误差分析；

⑦结论；

⑧思考题。

1.7 有机化学的文献资料

有机化学的文献资料非常丰富，现作简单介绍。

（1）工具书

① CRC Handbook of Chemistry and Physics　美国（CRC Press，Boca Raton，FL）出版的一本化学与物理英文手册。该书内容分六个方面：数学用表、元素和无机化合物、有机化合物、普通化学、普通物理常数和其他。在“有机化学”部分包含约 15000 个有机化合物及它们的物理性质，并按照有机化合物英文名字的字母顺序排列。查阅时可以根据英文名称，也可以根据该书提供的分子式索引（Formula Index）进行查找。

② Merck Index of Chemicals and Drugs（12th）　该书由美国（Merck and Co.；Rahway，NJ）出版的一本辞典，它主要介绍了有机化合物和药物，共收集了1万余种化合物的物理常数、制备、用途、毒性及参考文献等，并提供分子式索引和主题索引。

③ Beilstein Handbook of Organic Chemistry　该书内容较全，提供了有关化合物的来源、制备方法、物理性质、化学性质、生理作用、用途和分析方法等，并附有原始文献资料

的出处。该书可根据其编排原则进行查找，也可根据分子式索引或主题索引进行查找。

④ 化工辞典（4th）　综合性的化学工具书，收集了有关化学、化工名词 16000 余条，列出了物质名称的分子式、结构式、基本的物理化学性质和有关数据，并有简要的制法和用途说明。

（2）期刊

① 中文期刊　重要的有《中国科学》、《化学学报》、《化学通报》、《高等学校化学学报》、《有机化学》等。

② 英文期刊　重要的有《Journal of the Chemical Society（Perkin Transactions）》、《Journal of American Chemical Society》、《Journal of Organic Chemistry》、《Chemical Reviews》、《Tetrahedron》、《Tetrahedron Letter》、《Journal of Chemical Education》等。

（3）化学文摘　化学文摘是将大量的、分散的各种文字的文献加以收集、摘录、分类整理而出版的一种期刊。在众多的文摘性刊物中以美国化学文摘（Chemical Abstracts，CA）最重要。CA 创刊于 1907 年，现在每年出两卷，每周一期。CA 的索引系统比较完善，有期索引、卷索引，每十卷有累积索引。卷索引和累积索引主要有分子式索引（Formula Index）、化学物质索引（Chemical Substance Index）、普通主题索引（General Subject Index）、作者索引（Author Index）、专利索引（Patent Index）等。

（4）网上资源　由于互联网技术的迅速发展，从网上查找有关资料变得非常方便、迅速。从网上获取化学信息的途径很多，这里只作简单的介绍。

① 网上图书馆　Internet 上的图书馆是获取图书、杂志资料的重要途径之一。例如：
中国国家图书馆 http：//www. nlc. gov. cn/；
清华大学图书馆 http：//www. lib. qinghua，edu. cn/；
北京大学图书馆 http：//www. lib. pku. edu. cn/html 等。

② 中国期刊网　http：//www，china journal，net/；中文科技期刊数据库 http. //www，cqvip，com/。

③ 专利文献　IBM Intellectual Property Network：http//www. patents. lbm. com；例如：
中国专利信息网 http//www. patent. com. cn；
中国专利网 http//www. cnpatent. com。

④ 数据库资源　有关化学信息数据库中，化学结构数据库占有很高的比例，但也不乏一些范围较小的专业数据库，例如：
有机化合物数据库 http//www. colby. edu/chemistry/cmp/cmp. html；
化合物基本性质数据库 http//www. chemfinder. camsoft. com/；
上海有机化学研究所的红外光谱数据库 http：//www，sioc. ac. cn/lccdeb/sjk，htm（IR spectrum database）等。

⑤ 网络中有些资源可免费查阅，而有些资源的使用则需收费。例如：
http//www. acs. org/（美国化学会）
http//www. ccs. ac. cn/（中国化学会）
http//www. chemsoc. org/（ChemSoc，英国皇家化学会）
http//www. csir. org/（Chemistry Software and Information Resources）
http//www. organicchem. com/（有机化学网）
http//chemnet. com. cn/（中国化工网）
http//www. reagent. com. cn/（中国试剂网）
http//www. chemonline. net/（化学之门）

第2章　有机化学实验基本操作

2.1　加热和冷却

加热与冷却是有机化学实验中常用的两项基本操作技能。

2.1.1　加热

有机化学实验常用的热源有煤气、酒精和电能，加热方式有直接加热和间接加热。为避免直接加热可能带来的问题，根据实际情况可选用以下间接加热的方式。

(1) 空气浴加热　利用热空气间接加热的原理，对沸点在80℃以上的液体均可采用。实验中常用的方法有石棉网上加热和电热套加热。

(2) 水浴加热　加热温度在80℃以下，可用水浴加热。将容器浸于水中，使水的液面高于容器内液面。

(3) 油浴　油浴的加热范围为100～250℃，油浴所能达到的最高温度取决于所用油的品种。实验室常用的油有植物油（200～220℃）、液体石蜡（220℃）等。油浴加热要注意安全，防止着火，防止溅入水滴。

(4) 沙浴　当要求加热温度较高时，采用沙浴。沙浴温度可达350℃。

2.1.2　冷却

冷却是有机化学实验要求在低温下进行的一种常用方法，根据不同的要求，可选用不同的冷却方法。

一般情况下的冷却，可将盛有反应物的容器浸在冷水中；在室温以下冷却，可选用冰或冰水混合物；在0℃以下冷却，可用碎冰和某些无机盐按一定比例混合作为冷却剂。干冰（固体二氧化碳）和丙酮、氯仿等溶剂混合，可冷到−78℃；液氮可冷到−188℃。

必须注意的是当温度低于−38℃时，不能使用水银温度计，而使用装有有机液体的低温温度计。

2.2　干燥与干燥剂

干燥是有机化学实验中非常普通且十分重要的基本操作。干燥的方法大致有物理方法和化学方法两种。物理方法主要有吸附、分子筛脱水等；化学方法是用干燥剂去水，根据去水原理不同可分为：与水结合生成水合物和与水起化学反应。

2.2.1　液体有机物的干燥

2.2.1.1　干燥剂的选择

液体有机物的干燥通常是将干燥剂直接放入有机物中，因此选择干燥剂要考虑以下四点：与被干燥的有机物不能发生化学反应；不能溶于该有机物中；吸水量大，干燥速度快；价格便宜。表2.1列出常用干燥剂的性质。

表 2.1　常用干燥剂的性质

干燥剂	吸水容量	强度	速度	干燥剂	吸水容量	强度	速度
$CaCl_2$	0.97($CaCl_2 \cdot 6H_2O$)	中	快且持久	K_2CO_3	0.2	极弱	慢
$MgSO_4$	1.05($MgSO_4 \cdot 7H_2O$)	弱	很快	CaO	—	强	很快
Na_2SO_4	1.25	弱	慢	P_2O_5	—	强	快
Ca_2SO_4	0.06	强	快	分子筛	约 0.2	强	快

2.2.1.2　干燥剂的用量

干燥剂的用量是根据干燥剂的吸水量和水在有机物中的溶解度来估算，一般用量比理论值高。当然也要考虑分子结构，含亲水性基团的化合物用量稍多些。干燥剂的用量要适当，用量少干燥不完全；用量过多，则会因干燥剂表面吸附而造成被干燥有机物的损失。一般用量为 10mL 液体约加 0.5～1g 干燥剂，各类有机物常用干燥剂参见表 2.2。

表 2.2　各类有机物常用干燥剂

有机物	干燥剂	有机物	干燥剂
烃	$CaCl_2$，Na，P_2O_5	酮	K_2CO_3，$CaCl_2$，$MgSO_4$，Na_2SO_4
卤代烃	$CaCl_2$，$MgSO_4$，Na_2SO_4，P_2O_5	酸，酚	$MgSO_4$，Na_2SO_4
醇	K_2CO_3，$MgSO_4$，CaO，Na_2SO_4	酯	K_2CO_3，$MgSO_4$，Na_2SO_4，K_2CO_3
醚	$CaCl_2$，Na，P_2O_5	胺	KOH，NaOH，K_2CO_3，CaO
醛	$MgSO_4$，Na_2SO_4	硝基化合物	$CaCl_2$，$MgSO_4$，Na_2SO_4

2.2.1.3　操作方法

干燥前要尽可能地把有机物中的水分除去，加入干燥剂后，振荡，静置观察，若干燥剂黏附在瓶壁上，则应再加些干燥剂。干燥前呈浑浊，干燥后为澄清，可认为水分基本除去。反之，则说明水分太多，要先除水。干燥剂颗粒大小要适当，太大吸水慢；太小吸附有机物较多。

2.2.2　固体有机物的干燥

晾干：将固体样品放在干燥的表面皿或滤纸上，摊开，再用一张滤纸覆盖，放在空气中晾干。

烘干：将固体样品置于表面皿中，放在水浴上烘干，也可用红外灯或烘箱烘干。必须注意样品不能遇热分解，加热温度要低于样品的熔点。

其他干燥方法：干燥器干燥，减压恒温干燥器干燥等。

2.3　过滤

过滤法是利用多孔介质（如滤纸、滤布）截留固液悬浮液中的固体颗粒，而实现固液完全分离的方法。常用的过滤方法有常压过滤、减压过滤、热过滤等，固体物质就留在滤纸上，液体则通过过滤器滤入接收的容器中，过滤所得的溶液叫滤液。

固液混合物的温度、黏度、过滤时的压力和物质的状态都会影响过滤速度。热的溶液比冷的溶液容易过滤；溶液的黏度越大，过滤速度越慢；减压过滤比常压过滤快；固体物质呈胶状时必须先加热一段时间来破坏它，否则它要透过滤纸，总之，要综合各方面的因素选用不同的过滤方法。

2.3.1 常压过滤

常压过滤法是指在常压下用普通漏斗过滤的方法，漏斗的上口与下口处均为常压，过滤的动力为重力。此法最为简便和常用，但过滤速度较慢。

根据沉淀的性质选择滤纸的类型：细晶形沉淀选择慢速滤纸，胶体沉淀选择快速滤纸，粗晶形沉淀选择中速滤纸。

先把滤纸折叠成四层，展开呈圆锥形，将之放入圆锥形漏斗中，且要保证滤纸的角度与漏斗的角度一致，两者间不可有气泡；漏斗颈内要形成水柱，以便借助液柱的重力加快过滤速度；还要注意“三低、三靠”，以保证无滤液损失，如图 2.1 所示。

此法常用于除去不溶性物质，对于颗粒较大、吸水性差的晶体类产品有时也可用此法分离出晶体，分析化学中也用此法获得沉淀物。

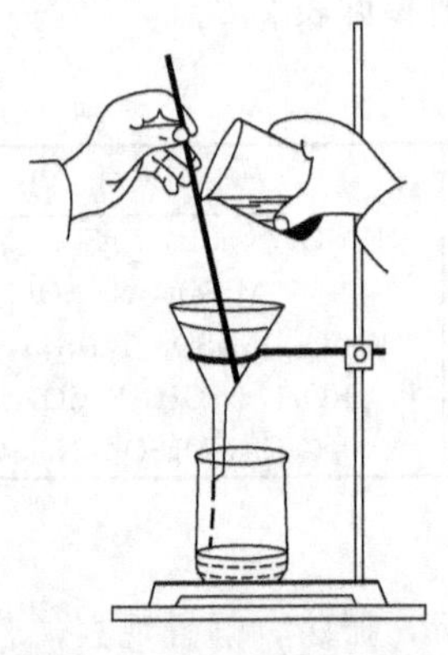

图 2.1 常压过滤装置

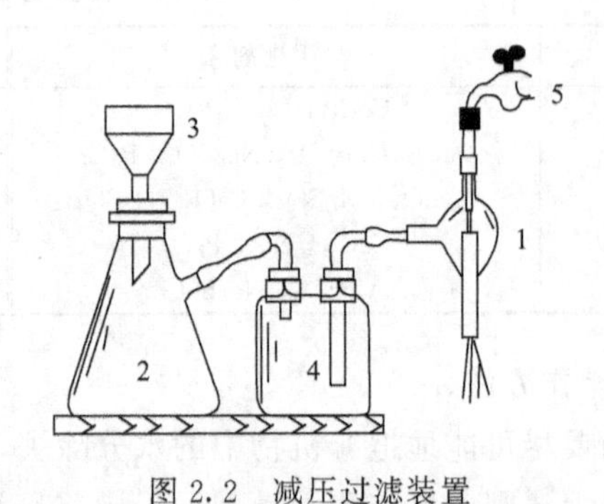

图 2.2 减压过滤装置

1—水泵；2—吸滤瓶；3—布氏漏斗；4—安全瓶；5—水龙头

2.3.2 减压过滤

减压过滤是利用泵急速把吸滤瓶内的空气带走，使布氏漏斗内液面与吸滤瓶内造成一个压力差，从而提高了过滤速度。在泵与吸滤瓶之间安装一个安全瓶，用以防止因关闭泵引起的倒吸现象。如图 2.2 所示。

使用此法过滤时要注意滤纸应比布氏漏斗内径略小，但又能把瓷孔全部盖住。放入滤纸后用蒸馏水润湿后打开泵，待滤纸贴紧滤板后再转移样品。抽滤结束后，应先打开安全泵上的活塞再关泵。

此法不适合胶状沉淀和颗粒太细的沉淀的过滤。

2.3.3 热过滤

如果物质在溶剂中的溶解度随温度变化较明显，为防止过滤时因温度降低而使晶体在滤纸上析出，就要趁热进行过滤，过滤时把玻璃漏斗放在铜质的保温漏斗内，保温漏斗内装有热水，以维持溶液的温度，装置见图 2.3。

过滤前可把玻璃漏斗放在水浴上用蒸汽加热，然后使用，且玻璃漏斗的颈部越短越好，以免过滤时溶液在玻璃漏斗颈内停留过久，因散热降温、析出晶体而发生堵塞。过滤用的滤纸折叠成褶皱滤纸（见图 2.4），以增大过滤面积、加快过滤速度。过滤时，溶液的转移要迅速，转移后漏斗上放一个凹面向下的表面皿，以防止溶剂的挥发。

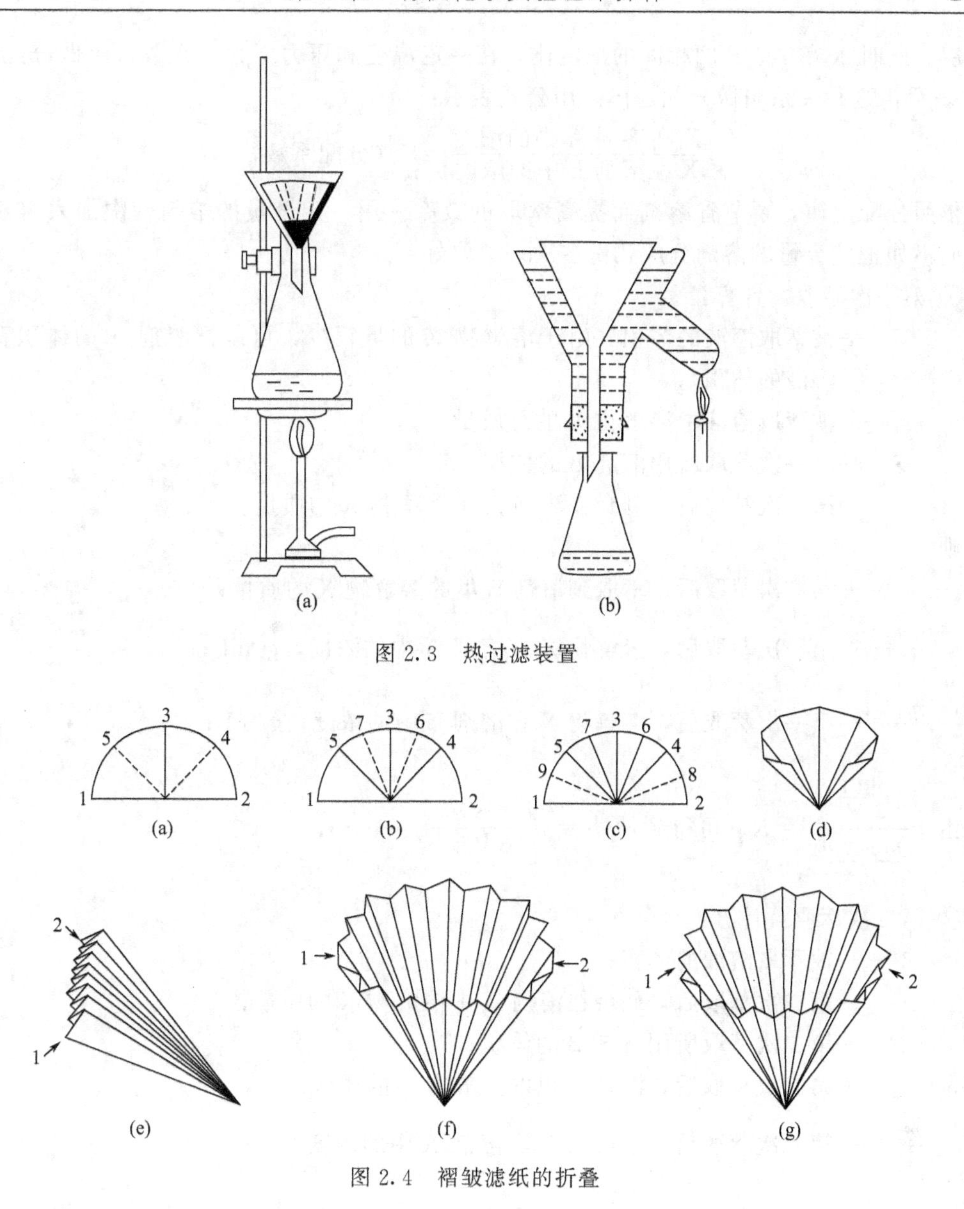

图 2.3 热过滤装置

图 2.4 褶皱滤纸的折叠

2.4 萃取和洗涤

2.4.1 原理

萃取是分离和提纯有机化合物常用的基本操作之一，用来提取的物质通常是固体或液体。提取液体的装置叫做分液漏斗，它的使用是有机化学实验中的必要操作之一。洗涤是萃取法的相反过程。通常情况下，通过另外一种不相溶的溶剂来提取溶液中的杂质，而该溶剂对所要提取的物质溶解度很小。分液漏斗也是洗涤的必要仪器，并且它的使用方法和萃取相似。在这里我们将集中讨论萃取法。

提取原则遵守以下计算方法。设溶液由有机化合物 X 溶解于溶剂 A 而成，现如要从其中萃取 X，我们可选择一种对 X 溶解度极好，而与溶剂 A 不相混溶和不起化学作用的溶剂 B。把溶液放入分液漏斗中，加入溶剂 B，充分振荡。静置后，由于 A 与 B 不相混溶，故分成两层。此时X在A、B两相间的浓度比，在一定温度和压力下为一常数，叫做分配系数，以K表示，这种关系叫做分配定律。用公式表示：

$$\frac{\text{X在溶剂A中的浓度}}{\text{X在溶剂B中的浓度}}=K\text{（分配系数）}$$

依照分配定律，要节省溶剂而提高萃取的效率，用一定分量的溶剂一次加入溶液中萃取，而不如把这个量的溶剂分成几份多次来萃取好。

① 第一次萃取 计算：

V——被萃取溶液的体积（由于溶解物X的量很少，可以把溶剂A的体积看作是溶液的体积）；

m_0——被萃取溶液中溶解物X的总质量；

S——第一次萃取所用溶剂B的体积；

m_1——第一次萃取后，剩余在溶剂A中溶解物X的质量。

则

m_0-m_1——第一次萃取后，萃取到溶剂B里的溶解物X的质量；

$\frac{m_1}{V}$——第一次萃取后，溶解物X在溶剂A中的浓度，g/mL；

$\frac{m_0-m_1}{S}$——第一次萃取后，溶解物X在溶剂B中的浓度，g/mL；

由于$\frac{\frac{m_1}{V}}{\frac{m_0-m_1}{S}}=K$，可得到：$m_1=m_0\frac{KV}{KV+S}$

② 第二次萃取 计算：

V——被萃取溶液的体积；

m_2——第二次萃取后，剩余在溶剂A中溶解物X的质量；

S——第二次萃取所用溶剂B的体积；

m_1-m_2——第二次萃取后，溶剂B里的溶解物X的质量；

$\frac{m_2}{V}$——第二次萃取后，溶解物X在溶剂A中的浓度，g/mL；

$\frac{m_1-m_2}{S}$——第二次萃取后，溶解物X在溶剂B中的浓度，g/mL。

所以

$$\frac{\frac{m_2}{V}}{\frac{m_1-m_2}{S}}=K\text{，可得到 }m_2=m_1\frac{KV}{KV+S}$$

用m_0代替m_1，上述关系式可再被写成

$$m_2=m_0\left(\frac{KV}{KV+S}\right)^2$$

我们可以采用类似的方法多次萃取。假如每次用于萃取的溶剂B的体积是S，m_n是多次萃取后剩余在溶剂A中溶解物X的质量。我们可从下式中得到m_n，

$$m_n=m_0\left(\frac{KV}{KV+S}\right)^n$$

2.4.2　操作

2.4.2.1　液-液萃取

从液体中萃取（或洗涤）通常是在分液漏斗中进行的。常用的分液漏斗有球形、锥形和梨形三种，在有机化学实验中，分液漏斗主要用于：a. 分离两种不互溶且不反应的液体；b. 从溶液中萃取某种成分；c. 用水或碱或酸溶液洗涤某种产品；d. 用来滴加某种试剂（即代替滴液漏斗）。

分液漏斗的容积以被萃取液体积一倍以上为宜，使用前应事先用水检查塞子和活塞是否渗漏，确认不漏水时方可使用。

萃取（或洗涤）时，将分液漏斗放在铁架台上的铁圈中，关紧活塞，将待萃取的水溶液和萃取剂依次倒入分液漏斗中，塞紧顶塞。取下漏斗，用右手握住漏斗颈并用手掌顶住顶塞，左手握住漏斗活塞处，拇指压紧活塞，食指和中指分叉在活塞背面。把漏斗放平，前后小心摇振。开始时，摇振要慢，几次后，将漏斗的下口指向斜上方无人处，用左手的拇指和食指旋开活塞，从指向斜上方的下口释放出漏斗内的压力，也称“放气”，一般每摇振 2～3 次就要放气一次，以免顶塞被顶开而喷出液体，造成事故。在经过几次摇振放气后，把分液漏斗放回铁圈中，将顶塞上的小孔对准漏斗上的通气孔，静止 3～5min，待液体完全分层后，打开上面的顶塞，再缓缓旋开活塞，将下层液体自活塞放出，上层液体从分液漏斗的上口倒出。将下层溶液倒回分液漏斗中，加入新的萃取剂继续萃取。重复上述操作，萃取次数一般为 3～5 次（如图 2.5 所示）。

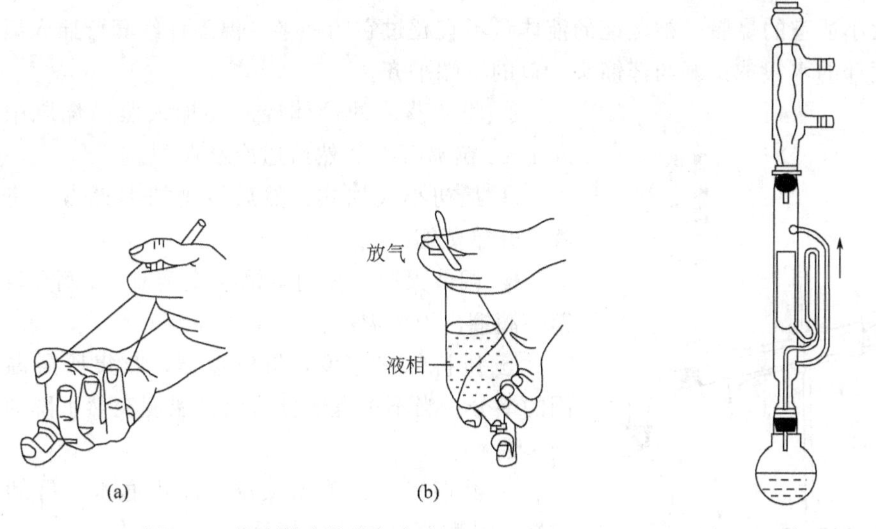

图 2.5　分液漏斗的排气方法　　图 2.6　索氏提取器

2.4.2.2　液-固萃取

自固体中萃取化合物，通常是用长期浸出法或采用脂肪提取器（如索氏提取器，见图 2.6），前者是靠溶剂长期的浸润溶解而将固体物质中的需要成分浸出来，效率低，溶剂量大。

脂肪提取器是利用溶剂回流和虹吸原理，使固体物质每一次都能被纯的溶剂所萃取，因而效率较高，为增加液体浸溶的面积，萃取前应先将物质研细，用滤纸套包好置于提取器中，提取器下端接盛有萃取剂的烧瓶，上端接冷凝管，当溶剂沸腾时，冷凝下来的溶剂滴入提取器中，待液面超过虹吸管上端后，即虹吸流回烧瓶，因而萃取出溶于溶剂的部分物质。就这样利用溶剂回流和虹吸作用，使固体中的可溶物质富集到烧瓶中，提取液浓缩后，将所得固体进一步提纯。

2.5 液体有机化合物的分离与纯化

2.5.1 蒸馏

2.5.1.1 原理

蒸馏是分离和纯化液体有机物常用的方法之一。当液体物质被加热时，该物质的蒸气压达到液体表面大气压时，液体沸腾，这时的温度称为沸点。常压蒸馏是将液体加热到沸腾状态，使该液体变成蒸气，又将蒸气冷凝后得到液体的过程。

每个液态的有机物在一定的压力下均有固定的沸点，利用蒸馏可将两种或两种以上沸点相差较大（>30℃）的液体混合物分开。但是应该注意，某些有机物往往能和其他组分形成二元或多元恒沸混合物，它们也有固定的沸点，因此具有固定沸点的液体，有时不一定是纯化合物。纯液体化合物的沸程一般为0.5～1℃，混合物的沸程则较长。可以利用蒸馏来测定液体化合物的沸点。

2.5.1.2 实验操作

① 按图2.7将实验装置按从下往上、从左到右的原则安装完毕，注意各磨口之间的连接。选一个大小适宜的烧瓶，被蒸馏的液体量不宜超过它的一半。温度计经套臂插入蒸馏头中，并使温度计的水银球正好与蒸馏头支口的下端平齐。

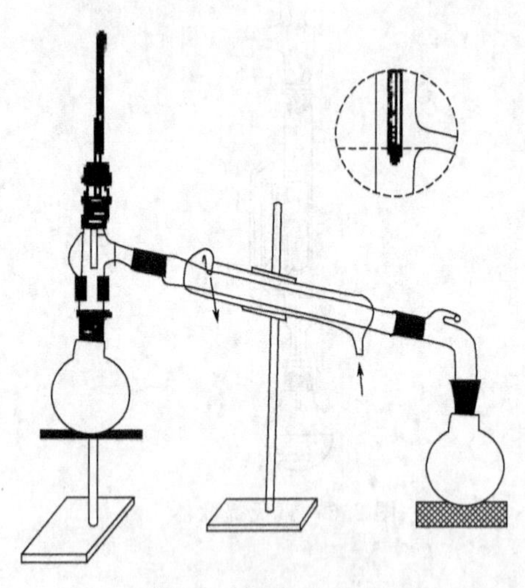

图2.7 常压蒸馏装置

② 将待蒸馏的液体经漏斗加入蒸馏烧瓶中，放入1～2粒沸石[1]，然后通冷凝水[2]。

③ 最初小火加热，然后慢慢加大火力，使之沸腾，开始蒸馏[3]。

④ 调节火源，控制蒸馏速度为1～2滴/s，记下第一滴馏出液的温度。

⑤ 维持加热速度，继续蒸馏，收集所需温度范围的馏分。当不再有馏分蒸出且温度突然下降时，停止蒸馏[4]。

⑥ 蒸馏完毕、关闭热源，停止通水，拆卸实验装置，其顺序与安装时相反。

注释：

[1] 加沸石可使液体平稳沸腾，防止液体过热产生暴沸；一旦停止加热后再蒸馏，应重新加入沸石；若忘记加沸石，应停止加热，冷却后，再补加沸石。

[2] 冷凝水从冷凝管支口的下口进，上口出。

[3] 此时的温度就是馏出液的沸点

[4] 切勿蒸干，以防止发生意外事故。

2.5.2　分馏

2.5.2.1　原理

分馏也是分离提纯液体有机物的一种方法。分馏主要适用于沸点相差不太大的液体有机物的分离提纯，其分离效果比蒸馏好。

分馏通常是在蒸馏的基础上用分馏柱来进行的。利用分馏柱进行分馏，实际上就是让在分馏柱内的混合物进行多次汽化和冷凝。当上升的蒸气与下降的冷凝液互相接触时，上升的蒸气部分冷凝放出热量使下降的冷凝液部分汽化，两者之间发生了热量交换，其结果是上升蒸气中易挥发组分增加，而下降的待凝液中高沸点组分增加。如果继续多次，就等于进行了多次的汽液平衡，即达到了多次蒸馏的效果。这样靠近分馏柱顶部易挥发物质的组分比率高，而烧瓶里高沸点组分的比率高。当分馏柱的效率足够高时，在分馏柱顶部出来的蒸气就接近于纯低沸点的组分，高沸点组分则留在烧瓶里；最终便可将沸点不同的物质分离出来。

2.5.2.2　实验操作

① 选一根分馏柱，按图 2.8 将实验装置安装完毕，注意各磨口之间的连接。

② 将待分馏液放入烧瓶中，加 1～2 粒沸石，通冷凝水。

③ 控制加热速度，使馏出速度为 1～2 滴/s[1]。

④ 分段收集馏分[2]，分别记下各馏分相应的温度，分别称量。

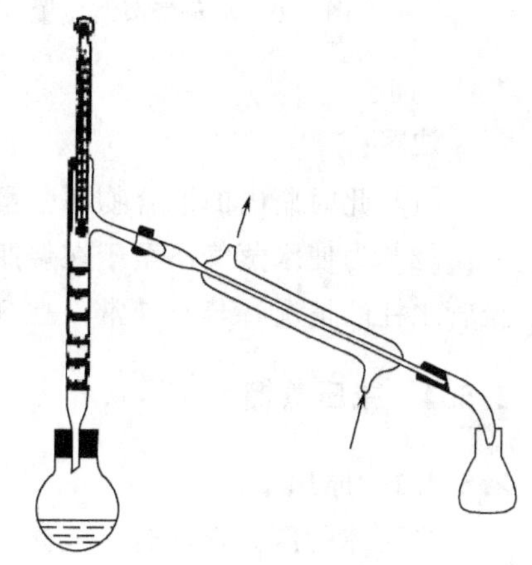

图 2.8　分馏装置

注释：

[1] 馏出速度太快，产物纯度下降；速度太慢，馏出温度容易上下波动。为减少热量散失，可用布将其包起来。

[2] 注意切不可蒸干。

2.5.3　水蒸气蒸馏

2.5.3.1　原理

水蒸气蒸馏也是分离提纯液体有机化合物的常用方法之一。它是将水蒸气通入不溶或难溶于水、有一定挥发性的有机物中，使该有机物随水蒸气一起蒸馏、冷凝，收集而得。

根据分压定律，混合物的蒸气压应该是各组分蒸气压之和。当各组分的蒸气压之和等于大气压力时，混合物开始沸腾。混合物的沸点要比单个物质的正常沸点低，这意味着该有机物可在比其正常沸点低的温度下被蒸馏出来。

在馏出物中，有机物与水的质量（m_A 和 m_{H_2O}）之比，等于两者的分压（p_A、p_{H_2O}）和两者各自分子质量（M_A 和 M_{H_2O}）的乘积之比

$$\frac{m_A}{m_{H_2O}}=\frac{p_A M_A}{p_{H_2O} M_{H_2O}}$$

水蒸气蒸馏常用在以下几种情况：①常压蒸馏易分解的高沸点有机物；②混合物中含大量固体，蒸馏、过滤、萃取等方法都不适用；③混合物中含有大量树脂状的物质或不挥发杂质，用蒸馏、萃取等方法难以分离。

若采用水蒸气蒸馏，被提纯物质应具备下列特点：①不溶于或难溶于水；②共沸腾下与

水不反应；③100℃时必须有一定的蒸气压。

2.5.3.2 实验操作

① 按图2.9安装实验装置，注意各磨口之间的连接，确保各连接点之间平顺，保持水蒸气发生器和三口瓶之间的连接尽可能短。它们之间用一个T形管连接，T形管下端连接一带有夹子的短橡皮管。

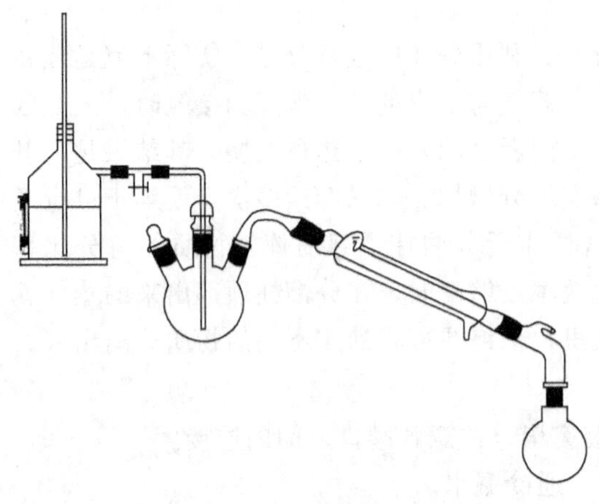

图2.9 水蒸气蒸馏装置

② 将蒸馏物倒入圆底烧瓶中，其用量不超过烧瓶容量的1/3，检查实验装置是否漏气。

③ 开始蒸馏前将T形管夹子打开，通冷凝水，加热水蒸气发生器，当T形管支管有蒸汽冲出时，夹紧夹子，使蒸汽通入烧瓶中[1]。

④ 调节热源，控制流出速度1～2滴/s[2]。

⑤ 当馏出物澄清透明时，即可停止蒸馏。

⑥ 先打开T形管的夹子，再移去火源，依次拆除装置。

注释：

[1] 此时瓶内的混合物不断翻滚，不久有机物和水的混合物馏出。

[2] 为使水蒸气不至于在烧瓶中过多而冷凝，可在烧瓶底下用小火加热。要随时注意安全管水柱的情况，若有异常，立刻打开T形管夹子，移去热源，排除故障后方可继续。

2.5.4 减压蒸馏

2.5.4.1 原理

当液体的压力降低时，液体可在低温下沸腾。因此，对沸点大于200℃的液体一般需用减压蒸馏提纯，高沸点有机化合物或在常压下蒸馏易发生化学反应的有机化合物，多采用减压蒸馏进行分离、提纯。

若系统的压力接近大气压时，每降低10mmHg（1.33kPa），物质的沸点降低0.5℃。低压时，压力降低一半，沸点可下降10℃。例如，某化合物在20mmHg（2.67kPa）的压力下，沸点为100℃，压力减至10mmHg（1.33kPa）时，沸点为90℃。

更精确一些的压力与沸点的关系可以用图2.10所示的沸点-压力关系图来估算。已知化合物在某一压力下的沸点便可近似地推算出该化合物在另一压力下的沸点。例如，已知某化合物在100kPa（760mmHg）的沸点234℃。估算2.0kPa（15mmHg）的沸点。用透明尺连接234℃（B线）到2.0kPa（15mmHg）（C线）交于A线（113℃），估算出2.0kPa（15mmHg）时沸点大约是113℃。

2.5.4.2 装置

图2.11是减压蒸馏的装置。主要装置包括热源、圆底烧瓶、克氏头、毛细管、温度计、冷凝管、减压接收管、接收瓶、夹子、压力表和减压泵等。

减压蒸馏和常压蒸馏之间的不同主要是在装置上，减压蒸馏装置系统是密封的，能够保

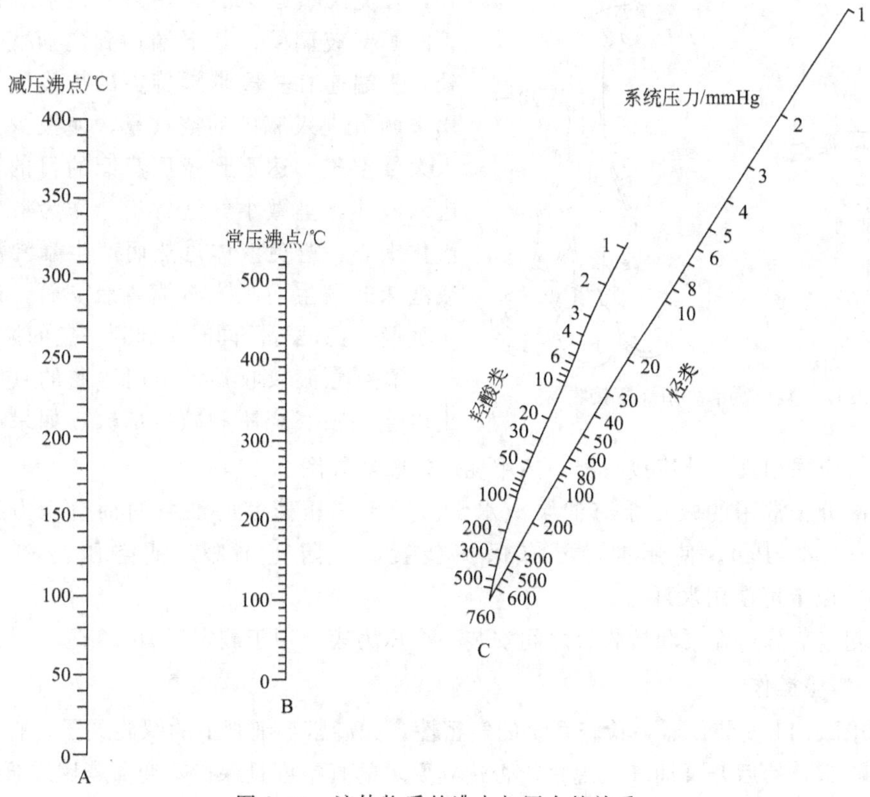

图 2.10　液体物质的沸点与压力的关系

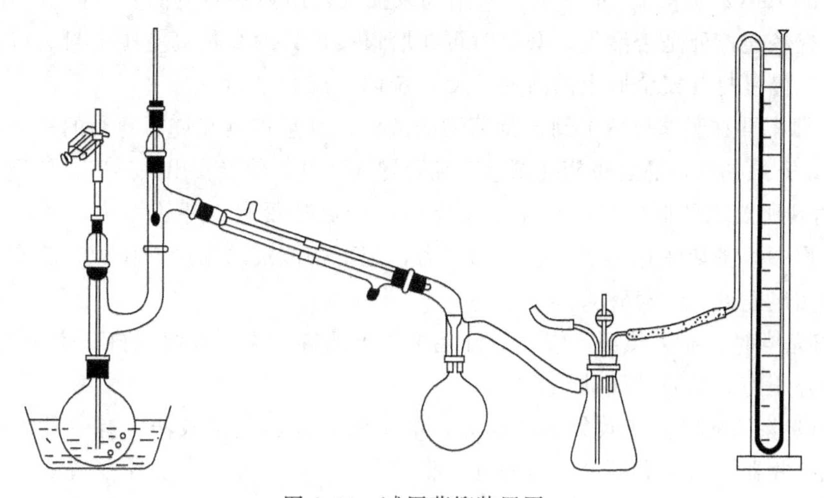

图 2.11　减压蒸馏装置图

持真空，所用的玻璃仪器必须能耐受减压，不能用大、薄壁的三角瓶，玻璃接口处要用真空脂密封。活塞不能涂抹太多的凡士林以免加热时不能保持真空。必须使用厚壁真空管，以防止减压时爆裂。连接的胶管和玻璃管，以及各接口处，必须是安全、密封的。

减压时，多数溶剂在室温下沸腾，因此，减压蒸馏能够除尽任何溶剂。

热源：热源应是能提供连续的、不变的热量，以防止加热不匀和过热，并能维持恒定的蒸馏速度。通常采用油浴。小规模的减压蒸馏可以用沙浴。

止沸装置：在减压过程中，可以用磁力搅拌器或微孔沸石来减少猛烈的沸腾和气泡。另

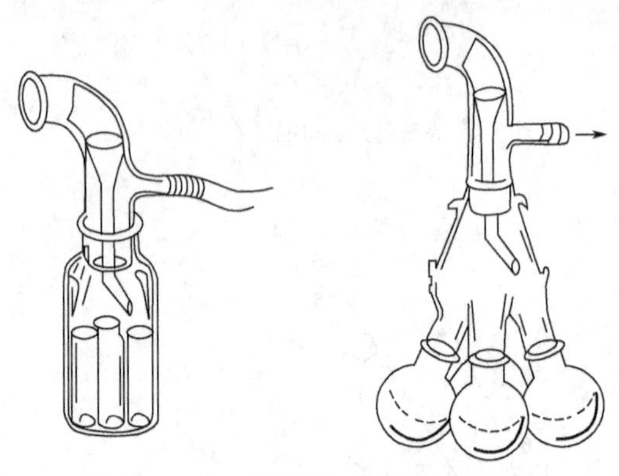

图 2.12 减压蒸馏多用烧瓶接收器

外，在克氏蒸馏头的一端插一根末端拉成毛细管的厚壁玻璃管，其下端伸到离瓶底 1～2mm 处，上端连有一段带螺旋夹的橡皮管。螺旋夹用来调节进入瓶中的空气量，使极少量的（引入大量空气，达不到减压蒸馏的目的）的空气进入液体，呈微小气泡冒出，作为液体沸腾的汽化中心，避免液体过热而产生暴沸溅跳现象或泡沫的发生而溅入蒸馏烧瓶支管，以保证减压蒸馏平稳进行。同时又起到一定的搅拌作用。

真空分流接收器：采用弯曲的真空接收管用以连接圆底烧瓶和减压部分，如果有多组馏分被收集，应采用真空分流接收器，多用烧瓶接收器如图 2.12 所示。

减压部分：常用的减压泵有油泵或水泵。用于减压蒸馏的油泵的抽空压力必须低于 20mmHg（2.67kPa）。使用时，要有保护装置，以防止倒吸。抽空压力在 20mmHg（2.67kPa）以上可使用水泵。

压力测定部分：在泵和接收管之间安装一个压力表，用于测定压力。

2.5.4.3 实验操作

① 按图 2.11 安装仪器，检查系统的气密性。先旋紧毛细管上的螺旋夹子，打开安全瓶上的二通旋塞，然后开泵抽气，观察能否达到要求的真空度且保持不变［若用水泵减压，一般可达 20mmHg（2.67kPa）的压力，若用油泵抽气，压力则会更低］。若发现有漏气现象，则需分段检查各连接处是否漏气，必要时可在磨口接口处涂少量真空脂密封。待系统无明显漏气现象时，慢慢打开安全瓶上的活塞，使系统内外压力平衡。

② 用长颈漏斗在蒸馏烧瓶中倒入待蒸馏的液体，其量控制在烧瓶容积的一半，关闭安全瓶上的活塞，开泵抽气，通过螺旋夹调节毛细管导入空气，使能冒出一连串小气泡为宜。

③ 当达到所要求的低压时，且压力稳定后，开启冷凝水，开始加热。采用油浴进行蒸馏。蒸馏过程中，密切注意蒸馏的温度和压力，记下待测液的温度和压力，若有不符，则应调节。控制馏出速度 1～2 滴/s。

④ 蒸馏完毕时，撤去热源，慢慢打开毛细管上的螺旋夹，并缓缓打开安全瓶上的活塞，平衡体系内外压力，然后关闭油泵。

⑤ 假如要除掉初馏分或收集多种馏分，必须断开泵转动接收器。有一种方便的方法是使用多烧瓶接收器，可不必断开泵接收几种馏分。

⑥ 移去接收烧瓶，拆除装置，清洗仪器。

2.5.5 旋转蒸发

旋转蒸发操作是分离有机溶剂常用的方法之一。旋转蒸发操作是在旋转蒸发器内完成的。旋转蒸发器的结构如图 2.13 所示，由一台电动机带动可旋转的蒸发器、冷凝管、接收瓶。它可在常压或减压下使用，可一次进料，也可分批进料。由于蒸发器在不断旋转，可免加沸石而不会暴沸。同时，液体附于壁上形成了一层液膜，加大了蒸发面积，使蒸发速度加快。停止蒸发时，应先停止加热，再停止抽真空，最后切断电源。

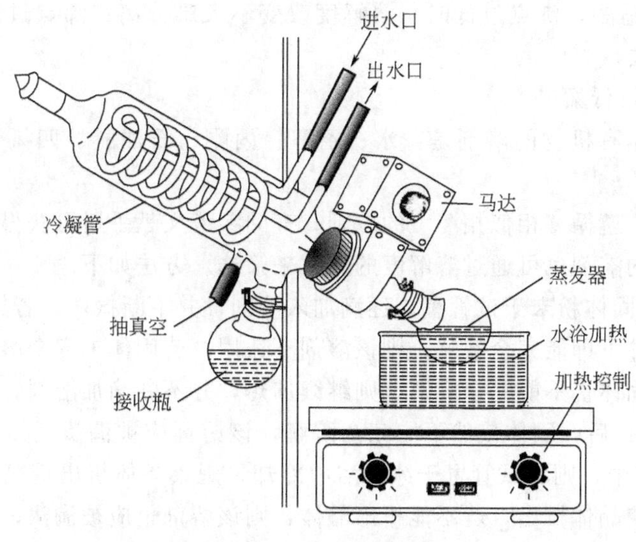

图 2.13　旋转蒸发装置

2.6　固体有机化合物的提纯

2.6.1　重结晶

2.6.1.1　原理

重结晶是利用混合物中各组分在某种溶剂中的溶解度不同或在同一溶剂中不同温度时的溶解度不同，而使它们相互分离的方法，它是提纯固体有机物常用的方法之一。

2.6.1.2　实验操作

（1）重结晶提纯法的一般操作

① 选择适宜的溶剂。

② 将粗产品溶于适宜的、热的溶剂中制成饱和溶液。

③ 趁热过滤除去不溶性杂质。如溶液的颜色深，则应先脱色，再行过滤。

④ 冷却溶液或蒸发溶剂，使之慢慢析出结晶而杂质留在母液中，或者杂质析出而欲提纯的化合物留在溶液中。

⑤ 抽气过滤分离母液，分出结晶体或杂质。

⑥ 洗涤结晶，除去附着的母液。

⑦ 干燥结晶。

（2）溶剂的选择　在重结晶操作中，溶剂的选择是提纯的关键。理想的溶剂具备下列条件：

① 不与被提纯物质起反应。

② 对被提纯物质的溶解度随温度变化较大，即，在较高温度时溶解度较大，而在较低温度时溶解度较小。

③ 对杂质的溶解度随温度变化较小，即，在较高温度和室温或较低温度时都很大或很小，前者是使杂质留在母液中不随待提取物晶体一起析出，后者是使杂质在趁热过滤时被滤去。

④ 能给出较好的结晶。

⑤ 溶剂的沸点适中，沸点过低时，溶解度改变不大难分离；沸点过高时，附着于晶体表面的溶剂不易除去。

⑥ 毒性小，价格低廉。

常用于提纯固体有机物的溶剂是：水、乙醇、丙酮、石油醚、四氯化碳、苯和乙酸乙酯等。

在选择溶剂时，遵循“相似相溶”的原则。查阅参考文献手册可获得一些常用物质的溶解度。当然，理想的溶剂也可通过溶解度的测试来筛选。方法如下。

称量大约 0.1g 固体粉末于试管里，逐滴加入溶剂，并不断振荡。若加入 1mL 溶剂后固体在室温或微热条件下即能完全溶解，则该溶剂不适用；若固体未完全溶解，可小心加热至溶剂沸腾，如果该固体仍不能完全溶解，则继续加热，并不断滴加溶剂，保持溶剂沸腾，若加入的溶剂量达 3mL 后，固体依然不能完全溶解，该溶剂应被淘汰。如果该溶剂能溶解在 1～3mL 沸腾的溶剂中，则将试管置于冰水浴中冷却，观察晶体析出情况，必要时可用玻璃棒摩擦试管壁，促使晶体析出，若不能析出晶体，则该溶剂也应被淘汰，如果晶体能正常析出，要注意观察晶体的析出量。在用同样的方法比较了几个溶剂后，就可选用晶体得率最好的溶剂来进行重结晶。如果没有适合的单一溶剂，可考虑使用两种溶剂的混合溶剂，其中一种溶剂对产品的溶解度很大，另一种对产品的溶解度很小，常用的混合溶剂有：乙醇和水，乙醇和乙醚，乙醇和丙酮，乙醚和石油醚，苯和石油醚等。

(3) 固体的溶解　按要求小心使用易燃溶剂。有机溶剂多数是易燃或有毒的，或二者兼而有之。切记要远离任何火源。化学反应要在通风橱中进行。常使用开口较细的三口瓶或圆底烧瓶作容器，以减少挥发，有利于搅拌下快速溶解。如果溶剂是低沸点且易燃应禁止在石棉网上直接加热。应根据溶剂的沸点，选择回流冷凝管安装在烧瓶上，并选择合适的热源，安装回流冷凝管可长时间加热并防止溶剂挥发。如图 2.14 所示。

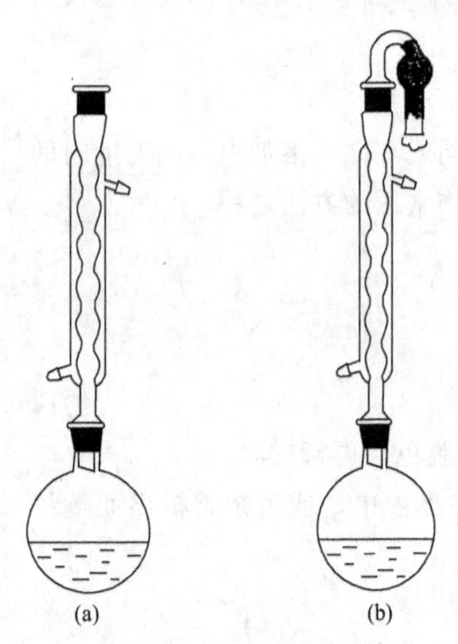

图 2.14　回流溶解装置

加粗产品到细口烧瓶中，在加热过程中滴加溶剂直至溶解。一般加入 20%～100%的溶剂，否则热过滤时晶体分离过快。假如溶剂量是未知的，可先加入少量溶剂并加热至沸点。假如固体仍未能完全溶解，可在沸腾状态下逐滴滴加溶剂直至溶解。

在溶解过程中，产品出现油状物则不利于重结晶，因为此时它会携带杂质和少量溶剂，应避免这种现象发生。可按下列方法操作：①选择的溶剂的沸点要低于溶质的沸点；②如果找不到合适的低沸点溶剂，可在低于溶质熔点的温度溶解该物质。

使用混合溶剂时，物质首先溶解在较高沸点的溶剂中。假如溶液是有色的，可加入活性炭除去有色杂质。热过滤时可滤出吸附有色杂质的活性炭，加热滤液接近沸点，滴加少量溶剂至产生浑浊。如果加热浑浊不消失，则应小心滴加一定量的溶剂至溶液澄清。

(4) 除去杂质　热过滤除去不溶性杂质和脱色用的活性炭。热过滤装置见图 2.3 及图 2.4。

(5) 晶体的分离　将热滤液放到一旁使其慢慢冷却几小时，或更长时间。既不要迅速冷却并搅拌得到细小的晶体，也不要缓慢冷却得到的结晶颗粒太大而不易干燥。当晶体颗粒长得太大时，可稍微摇动或搅拌几下，以使晶体颗粒大小趋于均匀。

如果滤液冷却后仍无晶体析出，可用少量晶种或摩擦器壁促使结晶形成。

如果晶体不能从某些油状物中分离，应加热溶液直至澄清，冷却溶液并搅动、分散直至油状物消失。

若利用结晶不能使物质纯化，可采用其他方法提纯（如层析，离子交换等）。

(6) 晶体的洗涤和收集　晶体的收集可通过瓷质的布氏漏斗的吸滤来获得。布氏漏斗配上胶塞安放在颈部连有胶管并连到泵上的厚壁吸滤瓶的上面，其装置见图 2.15。

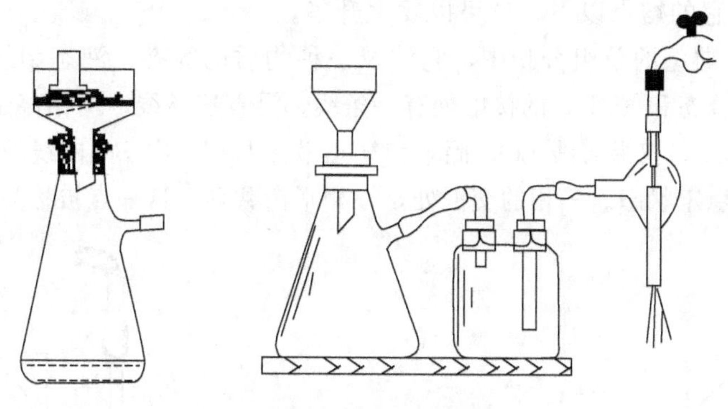

图 2.15　吸滤装置

在冲洗锥形瓶中的残留晶体时，应当用母液，不能用新溶剂，否则溶剂将晶体溶解，造成产品损失。当母液抽干后，布氏漏斗中的晶体要用溶剂洗涤，以除掉吸附在晶体表面上的母液。溶剂用量应尽可能少，以减少溶解损失。

洗涤的方法：先暂时停止抽气，在晶体上加少量溶剂，用刮铲或玻璃棒将晶体搅动，使所有晶体都被溶剂润湿，注意不要使滤纸松动。再开泵抽干溶剂。一般重复洗涤 1～2 次即可；晶体经抽滤、洗涤后，用刮铲将其转移至表面皿上进行干燥。

抽滤少量晶体时，可用玻璃钉漏斗，以吸滤管代替吸滤瓶。

(7) 晶体的干燥　抽滤、洗涤后的晶体，表面上还吸附有少量溶剂，因此还要用适当的方法进行干燥，将溶剂彻底除去。晶体的干燥方法有很多，常用的有晾干，烘干，用干燥器干燥等。

2.6.2　升华

2.6.2.1　原理

升华是提纯微量固体有机化合物的方法之一。

在样品中存在不挥发的杂质时，升华法是非常有效的。对于只需加热和手工操作又能保持其损失达到最低水平的不纯的固体，升华法是一种比较简单的方法。

升华法有很多优点：①它被选择用于净化高活性的物质，因为它在较高真空和较低温度下，非常有效；②最后痕量的溶剂都可以有效地被去除；③有低蒸气压的杂质更容易被分离出来，因为，这些理想的、有较低溶解度的物质在重结晶时很有可能被污染；④在实验过程中，溶解的物质易于溶解；⑤在以水作溶剂的情况下，升华法对于易溶解的物质是非常有效的。该法的缺陷是它不可能像重结晶法那样有选择性。这种选择性的缺陷在升华的物质蒸气压类似时发生。

升华法可在常压或减压下进行。减压时，在物质的熔点以下加热时，大量物质可以升华。某些分子间没有强烈作用力的物质通过升华被净化，如萘、二茂铁和对二氯苯是满足条件的化合物的例子。

当固体物质晶体被放加热升华时，通过蒸发过程逐渐产生气态分子（即，固体表现出蒸气压）。偶尔，气态分子中有的分子将撞击晶体表面或容器壁并通过分子间引力结合。后来的过程就是蒸发的反过程即冷凝。

许多固体物质在接近熔点时，有足够高的蒸气压，因此在实验室减压下，它们很容易升华。升华时，固体的蒸气压与外界的压力相等。

加热样品接近它的熔点以下，就可以发生升华。

图 2.16 是一个典型的单组分相图，它涉及一种物质的固态、液态和气态的温度和压力。两个区域（固态、液态和气态 ）的接触处有一条线，沿着这条线，两相存在平衡。线 *BO* 是物质的升华-蒸气压曲线，仅沿着线 *BO*，固态和气态平衡共存。在 *BO* 曲线的温度和压力下，液相是热力学上的不稳定状态。三相的交汇处，三相平衡共存。这一点被称为三相点。

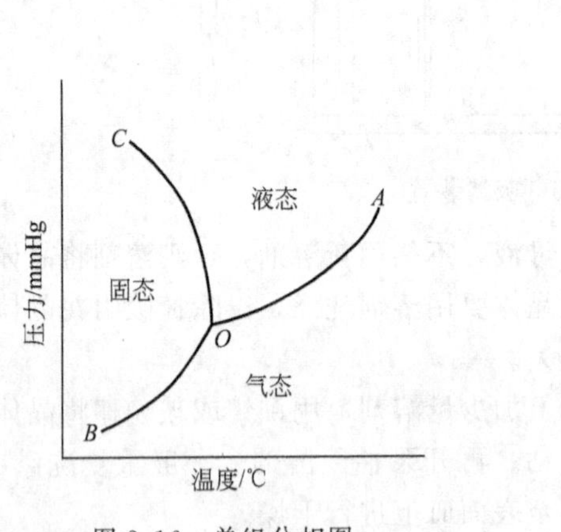

图 2.16　单组分相图

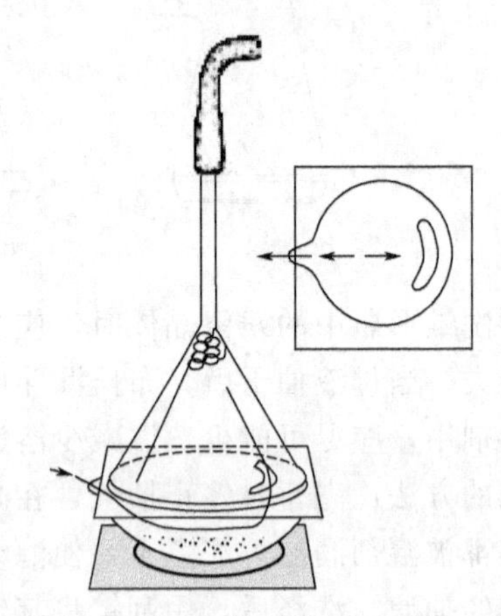

图 2.17　升华装置

在接近熔点时，许多固体物质有足够高的蒸气压，当固体的蒸气压等于提供的压力时可以升华。

在接近固体的熔点下，加热样品会使其升华。蒸气在冷管的表面冷凝，然而只有少量的挥发性残渣将留在烧瓶底部。目前，适合少量升华的装置在工业上是常用的。

图 2.17 是基于夹带原理的升华装置。穿过蒸发盘边缘的吸出的空气流动使物质升华，清除通过石棉网的气体而使它们冷凝在漏斗或玻璃棉上。空气流动必须足够缓和，升华才能不通过玻璃棉，而冷凝在漏斗内部或橡胶管内。

2.6.2.2　实验步骤

① 安装如上所述的升华的装置。

② 粉碎干燥的升华用样品，放到薄的升华容器中。安装仪器，打开抽气机或冷水泵。

③ 用合适的热源加热，直到升华开始。收集到冷凝器中。在没有熔化或升华时，要调节温度以得到适合的升华速率。如果冷凝器负载过多，则用冷却及拆卸装置的方法来除掉升华物，然后重新恢复并加热。

④ 当所有的化合物都升华或只剩一种不挥发的物质残留时，从热源上取下装置，使其冷却。

⑤ 仔细清除冷凝器，并用刀刮下晶体放入容器中。

2.7　色谱法

色谱法是分离、提纯和鉴定有机化合物的一种重要的物理方法。色谱法的基本原理是建立在相分配原理的基础上，因此，要将样品溶解在流动相中，流动相可以是液体，也可以是气体，然后流经固定相。固定相可以是色谱柱或是固体表面。各组分与固定相相互作用的时间足够长，它们就能被分开。

色谱法按照固定相和流动相可分为：①气-固色谱；②液-固色谱；③气-液色谱；④液-液色谱。固相可以使用硅胶等多种材料，液相可以在一些如硅胶、多孔玻璃珠或是聚合物等材料的载体上涂布固定液制成。①和②的原理是吸附作用，吸附色谱法是以分析物与固定相之间直接的相互吸附作用为基础。③和④的原理是分配作用，分配色谱与固定液相上存在的量有关。

液-固和液-液色谱可以在柱子和板上进行，所以前者亦称为柱色谱，后者通常有两种类型，一种是纸色谱，另外一种是薄层色谱。薄层色谱广泛应用于反应的监测和产物的制备。气-固色谱和气-液色谱只能使用柱色谱，通常称为气相色谱（GC）。

2.7.1　柱色谱

2.7.1.1　原理

柱色谱是分离、提纯复杂有机化合物的重要方法，也可用于分离量较大的有机物。

在柱色谱中，固定相是一种固体，用来分离流经它表面的具有选择性吸附力的液体的成分。产生吸附的相互作用的类型与产生分子间吸引作用是一样的，也就是，静电引力、络合作用、氢键和范德华力。

图 2.18 所示是一个分离混合物的柱色谱。这种色谱柱里充满了微小的、分散的固体颗粒，氧化铝或硅胶都可以作为固定相。作为被分离的混合物置于柱的顶部。如果混合物为固体，则必须被适当的溶剂溶解。首先，样品在柱的顶部被吸附，当作为流动相的洗脱剂流过柱子时，它将带有混合物的成分。由于固定相的选择性吸附力的作用，被分离的成分将以不同的速度向柱底移动。吸附力弱的化合物将比吸附力强的化合物更快地被洗脱，因为前者在流动相里有很高的百分比。逐步分离的成分示意见图 2.19。

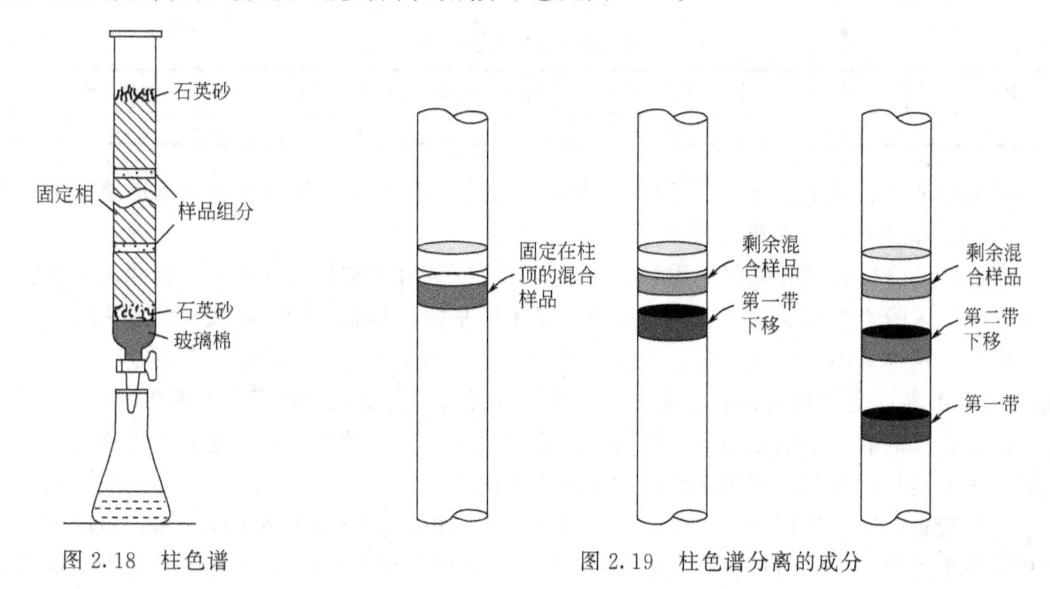

图 2.18　柱色谱

图 2.19　柱色谱分离的成分

2.7.1.2　吸附剂

在液相色谱中常用的固定相是硅胶和氧化铝。作为流动相的洗脱液是非极性或具有适中极性的溶剂。现今所用的吸附剂大多是硅胶，因为它所分离的样品的极性范围更广。同时，它本身也比氧化铝有优势，很少会与被分离的样品发生化学反应。氧化铝极性更强一些，因此，它对极性大的有机化合物的吸附力强，要用极性更强的溶剂才能进行洗脱。

色谱硅胶中有10%～20%的吸附水，在分配色谱中，它作为固定相时，化合物被洗脱剂和强吸收在硅胶表面的水所分离。

活性氧化铝商业上有售，有中性（pH＝7）、碱性（pH＝10）和酸性（pH＝4）三种，它们的吸附能力有很大的差异。

极性特强的化合物的分离需要吸附能力比硅胶和氧化铝还小的吸附剂。在此情况下使用的是反相色谱。反相色谱使用极性比流动相小的吸附剂为固定相，而极性大的样品先被洗脱出来，此时，极性小的样品在固定相上的吸附能力更强。

2.7.1.3　洗脱剂

在柱色谱中，洗脱剂一般具有一定的极性，与TLC类似，非极性化合物在固体吸附剂上的吸附力小，很容易用非极性的溶剂洗脱。极性化合物在金属氧化物上有很强的吸附力，因此，必须用极性更强的溶剂洗脱。

将非极性的洗脱液己烷与极性中等的乙醚进行对比，前者与氧化铝的作用是弱的范德华力，而后者与氧化铝的作用是偶极-偶极和配位作用。所以乙醚在氧化铝上的吸附力更强。

吸附剂的吸附力和溶剂的极性决定样品在柱子里的洗脱速度。样品洗脱得太快，分离效率就低，同样，吸附剂太强或溶剂极性太弱也是如此。相反，如果洗脱太慢，则需要选择有效的极性溶剂和温和的吸附。下面给出溶剂的洗脱能力顺序：

烷烃＜四氯化碳＜甲苯＜二氯甲烷＜乙醚＜氯仿＜丙酮＜乙酸乙酯＜乙醇＜甲醇

2.7.1.4　选柱和装柱

选好吸附剂后，还需确定其用量，这取决于被分离的样品和柱子的直径。样品在细长的柱子里比在短粗的柱子里保留的时间更长。一般来说，吸附剂的用量是样品用量的20～30倍。吸附剂高度与柱子直径之比是10∶1或15∶1。典型的色谱柱见表2.3。

表2.3　典型的色谱柱

样品/g	硅土/g	直径×柱长/cm	样品/g	硅土/g	直径×柱长/cm
5	150	3.5×40	2	60	2.0×30

氧化铝的密度是1g/cm³，而硅胶的密度大约是0.3g/cm³。例如20g氧化铝填充到直径为1.5cm的色谱柱时，高度达到11cm。

在选好色谱柱并称好吸附剂后，就可以装柱了。装好色谱柱与吸附剂、洗脱液的选择同样重要。如果吸附剂有裂缝或有流通渠道，吸附剂平面不规整，均会造成柱层析失败。

将色谱柱垂直固定在铁架台上，加入溶剂至柱子的1/2，用玻璃棉将下端塞住，再放入6mm的石英砂，塞子和石英砂用来支撑柱里的吸附剂，以防止它堵塞底部的尖端。

在充满溶剂的柱子里慢慢加入吸附剂粉末时，开关必须关闭。小心地将吸附剂置于柱子的底部，不能装得太紧，否则洗脱液的流速会很慢。

在把吸附剂加到溶剂中时，要轻轻敲打柱的边沿以防止吸附剂中出现气泡。吸附剂的顶部必须是水平的。加完后再小心加入4mm厚的石英砂。该砂层可以防止新加入溶剂对吸附剂平面的影响。

色谱柱一旦装好，不能干涸，而且溶剂液面不允许低于吸附剂的上层水平面。

2.7.1.5　洗脱技术

（1）溶解样品　加样时，通常把混合物溶解在溶剂中，然后加到柱子里。溶液尽可能浓缩，尽可能不超过 5mL。如果样品不溶解在开始用的洗脱溶剂中，可以用少量极性更大的溶剂进行溶解。

（2）加样品溶液　加样前让柱子里的溶剂流出，待溶剂液面刚好在石英砂上面时加入样品溶液，然后加入少量溶剂洗脱，等样品进入吸附剂后，将柱子加满洗脱液进行洗脱。

（3）洗脱化合物　极性很小的化合物用极性小的溶剂首先洗脱出来，然后换成极性大的溶剂洗脱有极性的化合物。在适当的条件下，随着洗脱的进行，不同的样品会分离成不同的色带。

有时也用混合溶剂进行洗脱。例如，开始用己烷洗脱时，没有样品流出，后改用含有 10％或 20％的乙醚-己烷溶液，必要时甚至换成纯的乙醚来进行洗脱。

可通过薄层色谱为柱色谱选择洗脱液，但是在使用相同吸附剂的情况下，对一个合适的洗脱剂而言，其在薄层色谱上 R_f 的取值范围是 0.2～0.8。

洗脱液的液面越高，流速越快。一般流速为 2mL/min。流速太快则不能达到吸附平衡，分离会不完全，但流速太慢会因为样品色带的自然扩散而使分离失败。

在柱子底部收集的每个部分的洗脱液的多少取决于特定的实验，每部分洗脱液的量是 25～150mL。如果分开的化合物有颜色，则收集不同色带的洗脱液；若为无色物质，应使用薄层色谱法跟踪鉴定各部分洗脱液，再合并相同组分的洗脱液。

（4）收集洗脱的化合物　蒸发洗脱的收集液，可重新得到纯的化合物。通常使用旋转真空蒸发仪。

2.7.1.6　柱色谱技术小结

① 准备一个合适的填装吸附剂的色谱柱；

② 向柱里的混合样品中添加少许溶液；

③ 用一定极性的溶剂来洗脱吸附的化合物；

④ 在柱的底部收集少许洗脱的化合物；

⑤ 蒸发溶剂，重新获得分离的化合物。

2.7.2　薄层色谱

2.7.2.1　原理

薄层色谱法是以薄层板作为载体，让样品溶液在薄层板上展开而达到分离的目的，故也称为薄层层析。它是快速分离和定性分析少量物质的一种广泛使用的实验技术，可用于精制样品、化合物鉴定、跟踪反应进程和柱色谱的先导（即为柱色谱摸索最佳条件）等方面。

薄层色谱法与柱色谱法原理相同且也是固-液吸附色谱。在薄层色谱里固定相以薄层（约 250μm）形式涂在玻璃板上。将分离的物质点在薄板上，然后把薄板放入盛有足够展开剂的层析缸中（不能没过点样线），混合物中的组分则以不同的速率随着溶剂展开。各成分的 R_f 值如图 2.20 所示。

薄层色谱技术非常容易和快捷，适用于分析混合物的成分，也可用于为以后的柱色谱优选最佳溶剂。但是必须记住，薄层色谱不适于分离沸点低于 100℃的挥发性化合物。

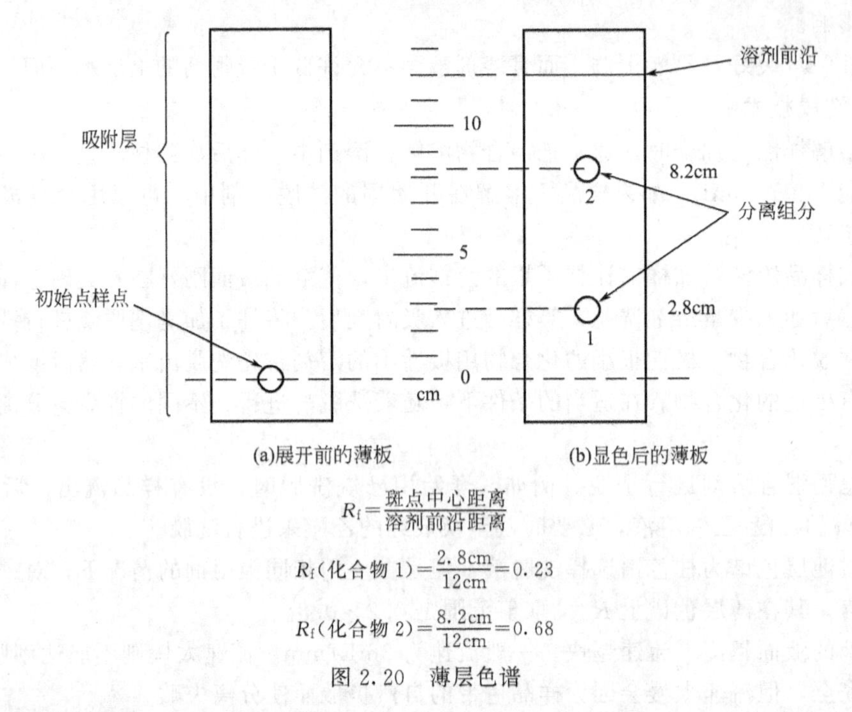

$$R_f=\frac{斑点中心距离}{溶剂前沿距离}$$

$$R_f(化合物1)=\frac{2.8\text{cm}}{12\text{cm}}=0.23$$

$$R_f(化合物2)=\frac{8.2\text{cm}}{12\text{cm}}=0.68$$

图 2.20 薄层色谱

2.7.2.2 薄层板材料

薄层板可以是玻璃、金属或塑胶的薄板。上面涂上一薄层吸附剂做固定相。

2.7.2.3 薄层色谱的展开和显色

在涂有吸附剂的薄板的一端，点上少量要分离的混合物。然后将薄板放入封闭的缸内。溶剂通过毛细作用浸过固定相。当溶剂前沿距板的顶端约 1cm 时，从层析缸中取出薄板。在溶剂挥干之前，立即用铅笔标出前沿的位置，可用两种方法显色：① 喷上荧光显色剂后用紫外灯照射；② 把薄板放到含有碘晶体的缸内，层析点显出碘的棕色。

2.7.2.4 R_f(比移值)的测定

R_f(比移值)表示物质移动的相对距离，即样品点到原点的距离和溶剂前沿到原点的距离之比（参见图 2.20）。R_f 值与化合物的结构、薄层板上的吸附剂、展开剂、显色方法和温度等因素有关。但在上述条件固定的情况下，R_f 值对每一种化合物来说是一个特定的数值。当两个化合物具有相同的 R_f 值时，在未做进一步的分析之前不能确定它们是不是同一个化合物。在这种情况下，简单的方法是使用不同的溶剂或混合溶剂来做进一步的检验。

2.7.2.5 薄层色谱常用的吸附剂

硅胶和氧化铝是薄层色谱常用的固相吸附剂。化合物极性越大，它在硅胶和氧化铝上的吸附力越强，所以吸附剂均制成活性精细粉末，活化通常是加热粉末以脱去水分。硅胶是酸性的，用来分离酸性或中性的化合物。氧化铝有酸性、中性和碱性的，可用于分离极性或非极性的化合物。商用的硅胶和氧化铝薄层板可以买到，这些薄板常用玻璃或塑料制成。溶剂在薄层板上爬升的距离越长，化合物的分离效果越好。宽的薄层板也可用于许多样品。具有 1～2mm 厚的大板可用于 50～1000mg 样品的分离制备。

2.7.2.6 样品的制备与点样

样品必须溶解在挥发性的有机溶剂中，浓度最好是 1%～2%。溶剂应具有较高的挥发性以便于立即蒸发。丙酮、二氯甲烷和氯仿是常用的有机溶剂。分析固体样品时，可将20～40mg 样品溶到 2mL 溶剂中。在距薄层板底端 1 cm 处，用铅笔画一条线，作为起点线，用毛细管（内径小于 1nm）吸取样品溶液，垂直地轻轻接触到薄层板的起点线上。样品量不能太多，否则易造成斑点过大、互相交叉或拖尾，不能得到很好的分离效果。

2.7.2.7　展开

将选择好的展开剂放在层析缸中，使层析缸内空气饱和，再将点好样品的薄层板放入层析缸中进行展开（参见图 2.21）。展开剂用量以使薄层板底部浸入溶剂 3～4mm 为宜。但溶剂不能太多，否则样点在液面以下，溶解到溶剂中，不能进行层析。当展开剂上升到薄层板的前沿（离顶端 5～10mm 处）或各组分已明显分开时，取出薄层板放平晾干，用铅笔画出前沿的位置后即可显色。根据 R_f 值的不同对各组分进行鉴别。

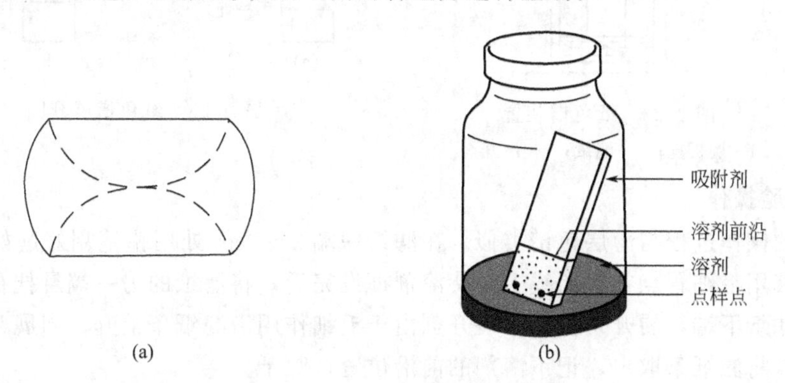

图 2.21　滤纸套（a）和薄层色谱层析缸（b）

2.7.2.8　显色

展开完毕，取出薄层板，画出前沿线，如果化合物本身有颜色，就可直接观察它的斑点；但是很多有机物本身无色，可先在紫外灯下观察有无荧光斑点。另外一种方法是将薄层板除去溶剂后，放在含有 0.5g 碘的密闭容器中显色来检查色点，许多化合物都能和碘形成黄棕色斑点。也可在溶剂蒸发前用显色剂喷雾显色。

2.7.2.9　薄层色谱技术小结

① 根据层析缸的大小选择已铺好的薄板；

② 用 1%～2%的含欲分离物质的溶液点样；

③ 用合适的展开剂展开；

④ 标出溶剂前沿；

⑤ 显色和圈出层析点；

⑥ 计算比移值 R_f。

2.7.3　纸色谱

2.7.3.1　原理

纸色谱属于分配色谱，是以滤纸作载体，让样品溶液在滤纸上展开达到分离的目的。纸色谱是用特制的滤纸作为惰性载体，以吸附在滤纸上的水或有机溶剂为固定相，流动相则是含有一定比例水的有机溶剂，通常称为展开剂。在滤纸的一定部位点上样品，当有机相沿滤纸流动经过原点时，即在滤纸上的水与流动相间连续发生多次分配，结果在流动相中具有较大溶解度的物质随溶剂移动的速度较快，而在水中溶解度较大的物质随溶剂移动的速度较慢，这样便达到混合物分离的目的。

常用的纸色谱装置如图 2.22 所示。

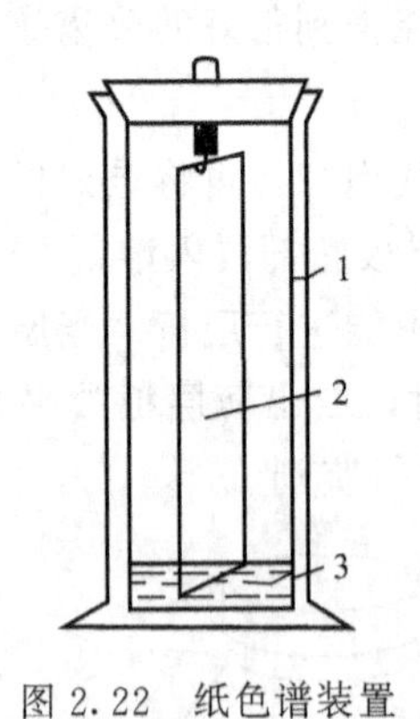

图 2.22　纸色谱装置

1—层析缸；2—滤纸；3—展开剂

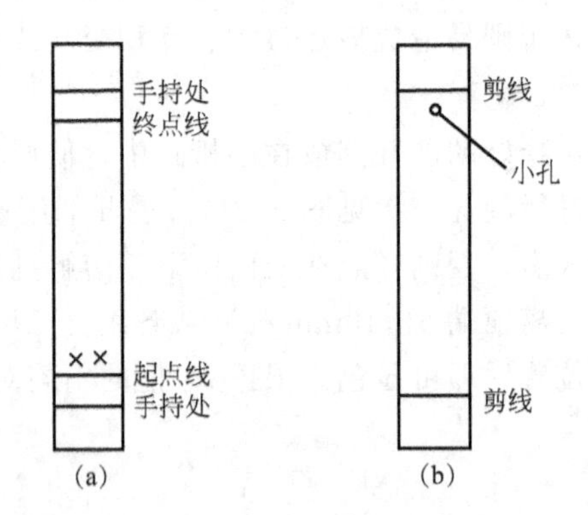

图 2.23　纸色谱展开图

2.7.3.2　实验操作

纸色谱的操作过程与薄层色谱类似。在滤纸一端 2～3cm 处用铅笔划好起始线，将被分离的样品溶液用毛细管点在起始线上，待溶剂挥发完后，将滤纸的另一端悬挂在层析缸的玻璃钩上，使滤纸下端与展开剂接触，展开剂由于毛细作用沿滤纸条上升。当展开剂前沿接近滤纸上端时，将滤纸条取出，记下溶剂的前沿位置，晾干。

若被分离的各组分是有色的，滤纸条上就有各种颜色的斑点显出，否则要用紫外光或显色剂显色。图 2.23 为纸色谱展开图。

计算化合物的 R_f。R_f 值与被分离化合物的结构、固定相与流动相的性质、温度以及滤纸的质量等因素有关。由于影响 R_f 值的因素很多，实验数据往往与文献记载不完全相同，因此在鉴定时常常采用标准样品作对照。

纸色谱法的优点是操作简单，价格便宜，色谱图可以长期保存；缺点是展开速度较慢。

2.8　有机化合物物理常数的测定

2.8.1　熔点的测定

2.8.1.1　原理

熔点是固体化合物在 101.325kPa 下固-液两相处于平衡时的温度。纯净的固体有机物一般都有固定的熔点，一个纯化合物从开始熔化（初熔）到完全熔化（全熔）的温度范围叫做熔程或熔距，其熔程一般不超过 0.5～1℃。当含有杂质时，熔点会有显著的变化，会使其熔点下降、熔程延长。因此，可以通过测定熔点来鉴定有机物，并根据熔程的长短来判断有机物的纯度。加热时，时间和温度改变时纯物质的相变如图 2.24 所示。

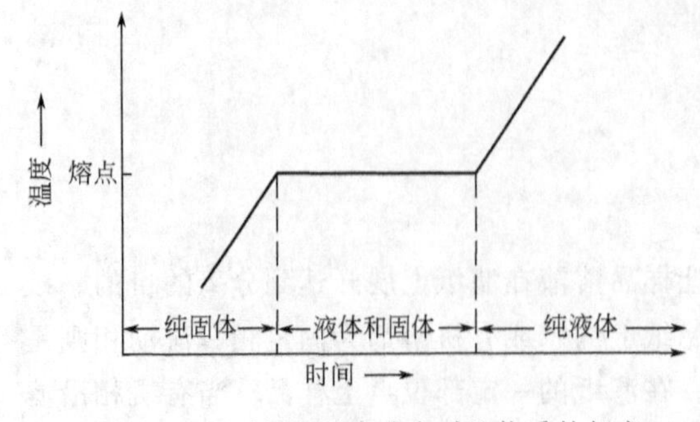

图 2.24　时间和温度改变时纯物质的相变

2.8.1.2　实验方法

方法一：毛细管法

（1）样品的填装　取少量干燥样品用研钵研细，堆成一小堆，将熔点管的开口端插入样品堆中，使样品挤入管内。然后把管开口一端向上，轻轻在桌子上墩几下，使样品掉入管底（参见图 2.25）。以同样方式重复取样几次。再取一支长约 45cm 的玻璃管垂直于表面皿上，将熔点管从玻璃管上端自由落下，重复多次，使样品装填紧密，高度约为 2～3mm。填装时操作要迅速，防止样品吸潮，装入的样品要结实。

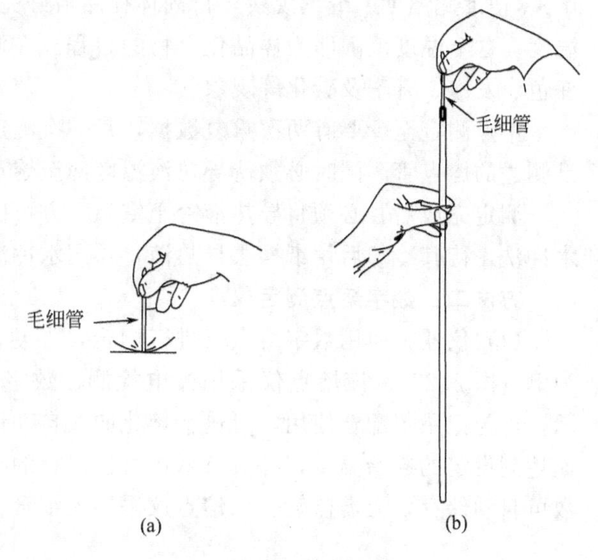

图 2.25　熔点管的装填

（2）仪器装置　毛细管法中最常用的仪器是 Thiele 管（又叫 b 形管或熔点测定管）。取一支 b 形管，固定在铁架台上，装入导热液（导热液一般用液体石蜡、硫酸或硅油等）至略高于支管口上沿。管口配一插有温度计的开槽塞子（也可将温度计悬挂），毛细管通过导热液紧附在温度计上，样品部分位于温度计水银球中部。并用橡皮圈将毛细管套在温度计上（橡皮圈不能浸入导热液中）。调整温度计位置，使其水银球恰好在 Thiele 管两侧管的中部（参见图 2.26）。

（3）熔点测定　测定时，先加热 Thiele 管，若测定已知样品的熔点，可先以较快速度加热。在距离熔点 5～20℃时，应控制加热速度，使温度每分钟上升 1～2℃至测出熔程；若测定未知样品要先粗测熔点范围，再用上述方法细测。

当毛细管中的样品开始塌落、并有小液滴出现时，表明样品已开始熔化即初熔（或始

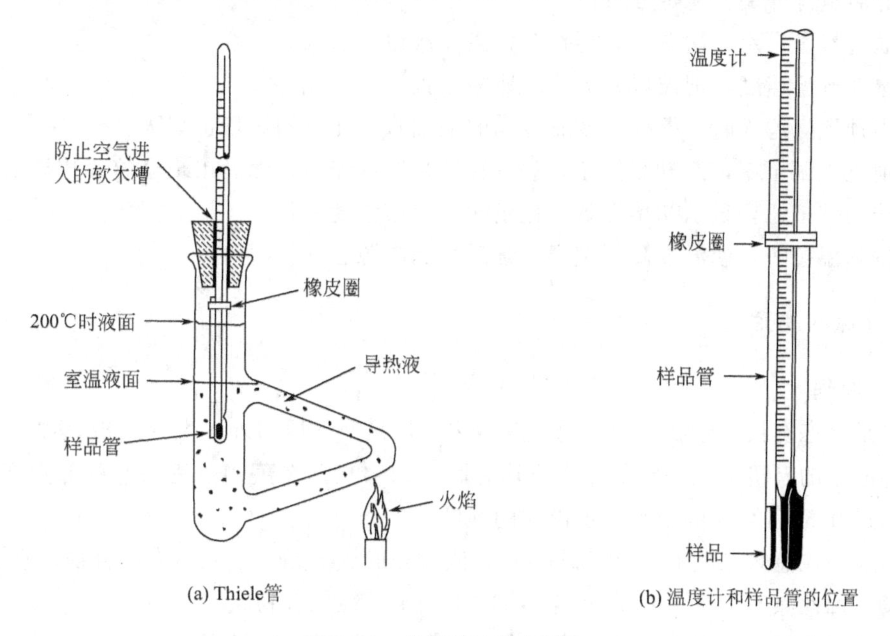

(a) Thiele管　　(b) 温度计和样品管的位置

图 2.26　毛细管法熔点测定

熔），记下此温度。继续观察，待固体样品恰好完全溶解成透明液体，即全熔时再迅速记下温度。这个温度范围即为样品化合物的熔程。在测定过程中，还要观察和记录是否有萎缩、变色、发泡、升华及碳化等现象。

熔点测定至少要有两次重复数据，每一次测定都必须用新的熔点管新装样品，不能使用已测过的熔点管。同时必须待导热液温度冷至熔点以下约15℃才能再进行测定。

测定完成后，必须将导热液冷至室温，方可倒回试剂瓶里。刚用完的温度计不能立即用水冲洗，待其冷却后用纸擦去导热液，再用水冲洗，以免温度计炸裂。

方法二：数字熔点测定仪

（1）仪器　使用数字熔点仪进行测定，方便、准确、易于操作。以WRS-1数字熔点仪为例（图2.27），该熔点仪采用光电检测、数字温度显示等技术，具有初熔、全熔自动显示，可与记录仪配合使用，可进行熔化曲线自动记录。该仪器采用集成化的电子线路，能快速达到设定的起始温度，并具有六挡可供选择的线性升、降温速率自动控制，初熔、全熔读数可自动储存，无需监管。该熔点仪采用毛细管作样品管。

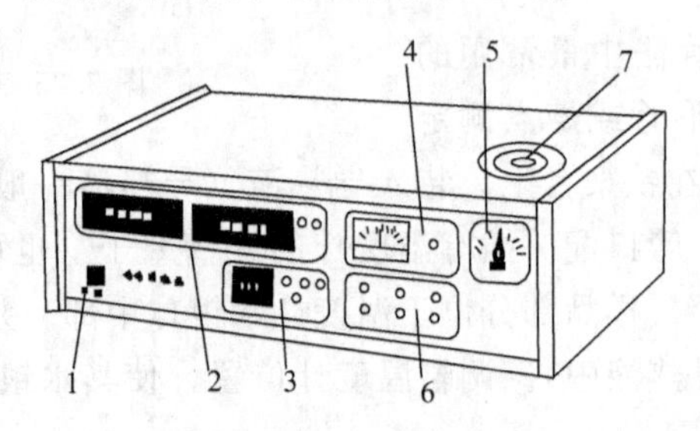

图2.27　WRS-1数字熔点仪

1—电源开关；2—温度单元；3—起始温度设定单元；4—调零单元；
5—速度选择单元；6—线性升、降温自动控制单元；7—毛细管插口

（2）操作步骤

① 开启电源开关，稳定20min。

② 通过拨盘设定起始温度，再按起始温度按钮，输入此温度。

③ 选择升温速度，把波段开关旋至所需位置。

④ 当预置灯熄灭时，可插入装有样品的毛细管，此时初熔灯也熄灭。

⑤ 把电表调至零，按升温按钮，数分钟后初熔灯先亮，然后出现全熔读数显示。

⑥ 按初熔按钮，显示初熔读数，记录初、全熔温度。

⑦ 按降温按钮，使温度降至室温，最后切断电源。

2.8.2　沸点的测定

2.8.2.1　原理

沸点是液态有机化合物的重要物理常数之一，是蒸气压与外界大气压相等时的温度。

每种液态的有机物在一定的压力下均有固定的沸点。在蒸馏过程中它的沸点变化很小，通常不超过0.5～1℃。可用来鉴定液体的纯度。

但是应该注意，某些有机物往往能和其他组分形成二元或三元恒沸混合物，它们也有固定的沸点，因此具有固定沸点的液体，有时不一定是纯化合物。

可以利用蒸馏来测定液体化合物的沸点。常压蒸馏的装置见图2.7。

2.8.2.2　实验操作

详见中文部分 2.5.1.2。

① 按图 2.7 安装实验装置。

② 将待蒸馏的液体经漏斗加入蒸馏烧瓶中，放入 1～2 粒沸石，然后通冷凝水。

③ 最初小火加热，然后慢慢加大火力，使之沸腾，开始蒸馏。

④ 调节火源，控制蒸馏速度为 1～2 滴/s，记下第一滴馏出液的温度（初沸点 t_1）。

⑤ 维持加热速度，继续蒸馏，收集所需温度范围的馏分。当不再有馏分蒸出且温度突然下降时，停止蒸馏。记下此时的温度（终沸点 t_2）。

⑥ 蒸馏完毕、关闭热源。停止通水，拆卸实验装置。

⑦ 整理数据：样品的沸程 t_1～t_2。

2.8.2.3　思考题

① 为什么不能将沸石直接加到热溶液中？

② 为什么常压蒸馏装置一端应该和大气相通？

2.8.3　折射率的测定

2.8.3.1　原理

折射率是液体化合物一个有用的物理常数，通过测定折射率可以判断有机化合物的纯度和鉴定未知物。折射率可精确到小数点后 4 位。

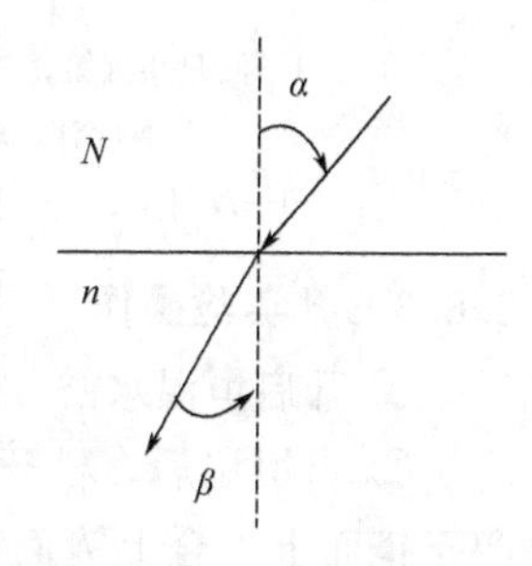

图 2.28　光的折射现象

光从一种介质进入另一种介质时，会发生折射，根据斯涅尔(Snellius)折射定律，波长一定的单色光，在确定的外界条件（如温度、压力等）下，从一种介质进入另一种介质时，入射角 α 与折射角 β 的正弦之比（见图 2.28）与这两种介质的折射率 N（光疏介质）和 n（光密介质）成反比，即：

$$\frac{\sin\alpha}{\sin\beta}=\frac{n}{N}$$

若光疏介质是真空，因真空的 $N=1$，则

$$n=\frac{\sin\alpha}{\sin\beta}$$

式中　n——一定温度、压力、波长条件下的折射率；

α——光疏介质中的入射角；

β——光密介质中的折射角。

所以一个介质的折射率，就是光线从真空进入这个介质时的入射角 α 与折射角 β 的正弦之比，这种折射率称为该介质的绝对折射率。通常测定的物质都是以空气为比较标准的，测得的折射率称为介质相对空气的折射率。折射率常受温度和光线的波长两个因素影响，所以表示折射率时必须注明测定时的温度和光波的波长。一般以钠光作为光源（波长为 589.3nm，以 D 表示），在 t℃时测定化合物的折射率，故折射率以 n_D^t 表示。温度升高，折射率减小。

当入射角 $\alpha=90°$时，折射角 β 达到最大值，称为临界角（β_0），则

$$n=\frac{1}{\sin\beta_0}$$

因此，通过测定临界角，就可以得到折射率。

测定液体的折射率，通常使用阿贝（Abbe）折光仪。

2.8.3.2 阿贝折光仪（Abbe）

测定液体化合物的折射率常用的仪器是阿贝折光仪，其结构见图 2.29。因具有如下优点而得到普遍应用：a. 能随镜像的调节直接读数；b. 用样量少；c. 样品的温度容易控制；d. 精确度高。

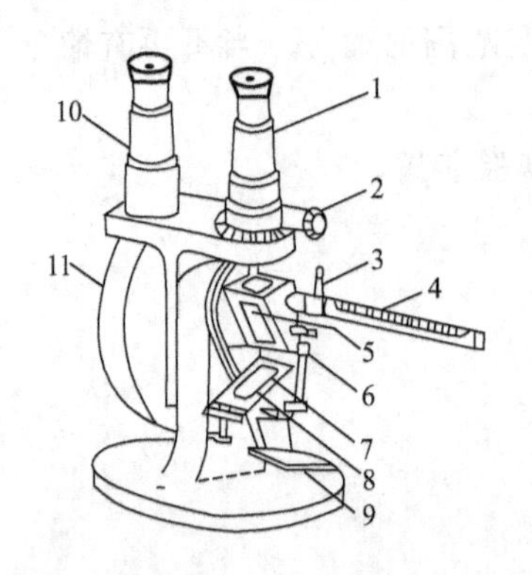

图 2.29　阿贝折光仪

1—测量望远镜；2—消色散手柄；3—恒温水入口；4—温度计；5～7—棱镜组；8—加液槽；9—反射镜；10—读数望远镜；11—折光仪刻度盘

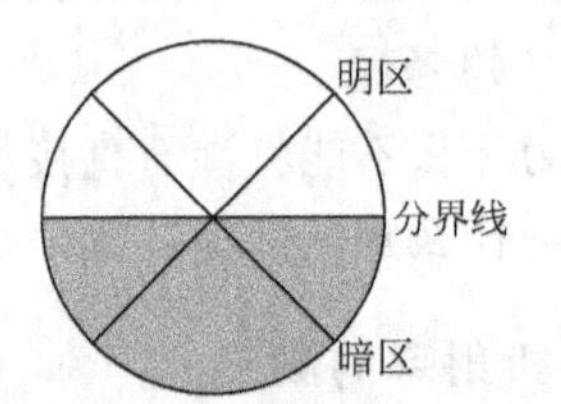

图 2.30　阿贝折光仪临界角时目镜视野图

2.8.3.3 实验操作

① 开启恒温水浴，通入恒温水（一般为 20℃或 25℃），若未连接恒温水浴，此步可免去。

② 当恒温后，松开锁钮，开启下面棱镜，使其镜面处于水平位置，滴 1～2 滴乙醇或甲醇于镜面上，合上镜面，促使难挥发的污物移走，再打开棱镜，用丝巾或擦镜纸轻轻擦拭镜面。

③ 轻轻打开棱镜，用滴管将 2～5 滴待测液均匀地滴在下面的磨砂面棱镜上[1]，要求液体无气泡并充满视场，关紧棱镜。若测定易挥发样品，可用滴管从棱镜间小槽滴入。

④ 调节反光镜和小反光镜，使两镜筒视场最亮。

⑤ 旋转棱镜转动手轮，使在目镜中观察到明暗分界线。若出现色散光带，可调节消色散棱镜手轮，使明暗清晰，然后再旋转棱镜转动手轮，使眼睛分界线恰好通过目镜中“×”字交叉点[2]，如图 2.30 所示。

⑥ 记录从镜筒中读取的折射率[3]，读至小数点后四位，同时记下温度。重复测定 2～3 次，取其平均值为样品的折射率。

⑦ 仪器用毕后用蘸有少量乙醇或乙醚的擦镜纸擦拭干挣，晾干两镜面，然后合紧两镜面，用仪器罩盖好或放入木箱内[4]。

注释：

[1] 操作时要特别小心，严禁触及棱镜，特别是油手、汗手及滴管的末端等。

[2] 若边界有颜色或出现漫射，可转动消色散棱镜即消色补偿器，直至边界呈无色和明暗界线清晰。

[3] 若使用数显折光仪，可直接从荧光屏上读取数据。

[4] 测定有毒样品的折射率时，应在通风橱内操作。

2.8.4 比旋光度的测定

2.8.4.1 原理

比旋光度像物质的熔点和沸点一样，是有机化合物的特征物理常数，可用来鉴定化合物及含量。

比旋光度［α］是标准状态下的旋光度α，可通过旋光度α计算获得。而旋光度可通过旋光仪测定。

旋光度是平面偏振光透过有机物的晶体或溶液时，其振动平面旋转的角度。旋光度与物质的性质、浓度、溶剂、温度、所用光线的波长及被透射物质的液层厚度等因素有关。旋光度通常用α_D^t表示，示为测定温度为t℃、钠光灯照射下的数值，比旋光度则用$[\alpha]_D^t$表示。

溶液的比旋光度：$[\alpha]_D^t=\dfrac{\alpha}{Lc}$

纯溶剂的比旋光度：$[\alpha]_D^t=\dfrac{\alpha}{Ld}$

式中　α——旋光度，(°)；

t——测定温度，℃；

D——钠光线（589.3nm）；

L——旋光管的长度，dm；

c——被测溶液的浓度，g/mL；

d——纯溶剂的密度，g/mL。

2.8.4.2 旋光仪及使用

图 2.31 为 WXG-4 圆盘旋光仪的简图，该仪器采用三分视野法来确定光学零位。

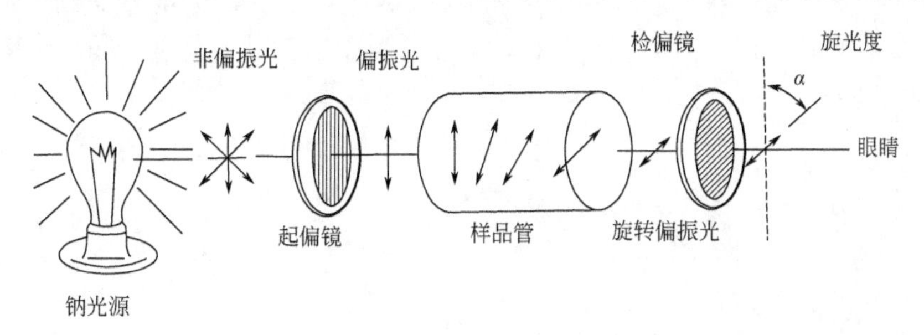

图 2.31　旋光仪简图

从钠光源射出的光线，通过聚光镜、滤色镜经起偏镜成为平面偏振光，在半波片处产生三分视场。通过检偏镜及物、目镜组可以观察到如图 2.32 所示的三种情况。转动检偏镜，在 0°时视场中三部分亮度一致，如图 2.32（a）。

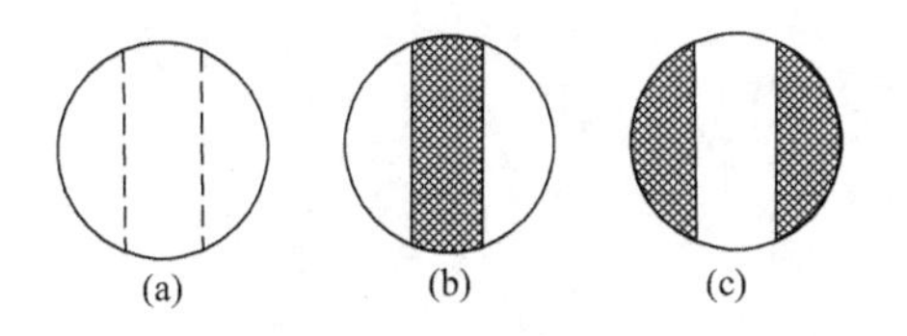

图 2.32　三分视场

当放入装有溶液的样品后，使平面偏振光旋转了一个角度，零度视场便发生了变化如图2.32（b）或（c）所示，转动检偏镜，能再一次出现亮度一致的视场，这个转动的角度就是溶液的旋光度，它的数值可通过放大镜从刻度盘读出。

2.8.4.3 **实验操作**

（1）预热　打开电源开关，预热5min。

（2）旋光仪的校正　用蒸馏水校正旋光仪的零点

① 装旋光管；

② 调节，测定；

③ 记录读数；

④ 重复测2～3次。

（3）样品溶液的旋光度的测定

① 已知浓度样品的旋光度的测定，复测2～3次；

② 未知浓度样品的旋光度的测定，复测2～3次。

（4）计算溶液的比旋光度。

第3章　综合实验

第一部分　制备

3.1　环己烯的制备

3.1.1　目的

① 学习利用环己醇在酸催化下脱水制备环己烯的原理和方法。

② 掌握分馏和简单蒸馏技术。

3.1.2　原理

环己烯可由环己醇在浓硫酸或浓磷酸催化下脱水制备，由于有浓硫酸时，反应物等容易炭化，本实验以浓磷酸为脱水剂制备环己烯。

$$\text{环己醇(OH)} \xrightarrow[\triangle]{H_3PO_4} \text{环己烯} + H_2O$$

3.1.3　实验仪器和主要试剂

(1) 实验仪器　分馏、蒸馏装置。

(2) 主要试剂　环己醇，浓磷酸，NaCl，Na_2CO_3，无水 $CaCl_2$。

3.1.4　实验步骤

① 在 50mL 圆底烧瓶中，加入 10.4mL 环己醇、4mL 85%的浓磷酸和几粒沸石，充分振荡使之混合。

② 安装分馏装置[1]（见英文部分图 1.7)，使用短分馏柱，用 50mL 锥形瓶作接收器，并置于冰浴中，以免环己烯挥发而损失。

③ 用石棉网加热反应混合物至沸腾，控制分馏柱顶部的温度不超过 90℃[2]，慢慢蒸出生成的环己烯和水。

④ 当烧瓶中只剩下少量的残留物时，即可停止加热，全部蒸馏时间约需 1h。

⑤ 将馏出液用 1g 固体氯化钠饱和，并轻轻地振摇烧瓶[3]。

⑥ 将馏出液倒入一分液漏斗中，加入 3～4mL 5%碳酸钠溶液（或用 0.5mL 20%的氢氧化钠溶液)[4]。轻轻振摇分液漏斗直至漏斗放气时无气体逸出。

⑦ 放出下层的水层，上层的粗产物从分液漏斗的上口倒入一个小锥形瓶中，加入 1～2g 无水氯化钙，振荡反应瓶，直至溶液变为澄清[5]。

⑧ 将干燥后的环己烯滤入一个干燥的 50mL 圆底烧瓶中，加入几粒沸石，用水浴加热蒸馏[6]，用一已称重的 50mL 圆底烧瓶作接收器，并将烧瓶置于冰水浴中。

⑨ 收集 80～85℃的馏分，称重产品，计算产率。

纯的环己烯为无色液体，b. p. 82. 98℃，d_D^{20} 0. 8102，n_D^{20} 1. 4465。

注释：

[1] 将生成的环己烯不断从反应体系中蒸馏出来，有利于反应平衡向产物方向移动。

[2] 环己烯的沸点为 83℃，环己烯与水共沸，沸点为 70. 8℃，原料环己醇也与水形成共沸物，沸点为 97. 8℃，因此，蒸馏温度需仔细控制，使其不超过 90℃。

[3] 加精盐的目的是减少产物在水中的溶解度，达到更好地分离的目的。

[4] 少量的磷酸也将与产物共沸，所以需用碳酸钠水溶液洗涤除去。

[5] 干燥后的溶液应澄清，若仍浑浊，可补加 0. 5g 干燥剂，振荡后再干燥 10min。

[6] 在精制产品时，蒸馏所用的仪器需在 110℃的烘箱内进行干燥，约 10min。

3. 1. 5 思考题

① 在反应期间，为什么必须控制分馏柱顶部的温度？

② 在液层被中和及分离前，加盐的目的是什么？

③ 产物环己烯被蒸馏前，为什么必须用碳酸钠将酸液调至中性？

3. 2 溴乙烷的制备

3. 2. 1 目的

掌握合成溴乙烷的原理和操作方法。

3. 2. 2 原理

制备溴乙烷的最简便的方法是将乙醇与溴化钠或溴化钾和硫酸混合物一起加热，溴化物与硫酸混合，放出溴化氢，溴化氢与乙醇反应生成溴乙烷。

主反应：

$$NaBr + H_2SO_4 \longrightarrow HBr + NaHSO_4$$

$$C_2H_5OH + HBr \rightleftharpoons C_2H_5Br + H_2O$$

溴化氢与乙醇的反应为可逆反应，为获得较高产率，必须促使平衡向生成产物的方向移动。在本反应中采用将溴乙烷从反应混合物中蒸出的办法，使平衡向产物方向移动。

因为反应中需要加入硫酸，故会引起以下副反应。

$$2C_2H_5OH \xrightarrow{H_2SO_4} C_2H_5OC_2H_5 + H_2O$$

$$C_2H_5OH \xrightarrow{H_2SO_4} CH_2{=}CH_2 + H_2O$$

$$2HBr + H_2SO_4(浓) \longrightarrow Br_2 + SO_2 + 2H_2O$$

伯醇发生取代反应时的温度一般不高，因此该反应的副反应不是很严重。

3. 2. 3 实验仪器和主要试剂

(1) 实验仪器　蒸馏、萃取装置。

(2) 主要试剂　溴化钠，95%乙醇，浓硫酸。

3.2.4　实验步骤

① 在100mL圆底烧瓶中加入10mL 95%乙醇及9mL水[1]，在不断振摇和冷水冷却下，慢慢加入19mL浓硫酸。

② 混合物冷至室温后，在冷却下加入12.9g研细的溴化钠，稍加振摇混合后，加入几粒沸石，按图1.7安装常压蒸馏装置。接收器内放入少量冰水，并将其置于冰水浴中，以防止产品的挥发损失。接引管的支管用橡皮管导入下水道或室外（为什么?）。

③ 通过石棉网用小火（或用加热套）加热烧瓶，使反应平稳进行[2]，直至无油状物馏出为止（约40min）。

④ 将馏出液倒入分液漏斗中，静置分层，将有机层（哪一层?）转移至一个干燥的锥形瓶中，并将其浸在冰水浴中冷却[3]，边振摇边滴加浓硫酸，以除去乙醚、乙醇、水等杂质，直至溶液有明显的分层为止。

⑤ 再用干燥的分液漏斗分去硫酸层（哪一层?）[4]。将溴乙烷粗产品转入干燥的蒸馏瓶中（如何转入?），加入沸石，水浴加热蒸馏。

⑥ 将已称量的、干燥的接收器外用冰水浴冷却，收集37～40℃的馏分，产率约为60.4%。

纯溴乙烷b.p.38.4℃，d_4^{20} 1.4604，n_D^{20} 1.4239。

注释：

[1] 加少量的水可以防止反应进行时产生大量泡沫，减少副产物乙醚的生成和避免氢溴酸的挥发。

[2] 反应开始时会产生大量的气泡，故应严格控制反应温度，使其平稳地进行，防止反应物冲出蒸馏瓶。溴乙烷沸点较低，蒸馏时一定要慢慢加热，否则蒸气来不及冷却而逸失，造成产品损失。

[3] 为防止产物挥发，在接收粗产品时接收器内外应用冰水冷却。

[4] 在产品纯化时，应尽可能将水分除净，否则当用浓硫酸洗涤时，由于放热，会使产品挥发损失。

3.2.5　思考题

① 造成本实验产率不高的主要原因是什么?

② 本实验根据哪种原料计算产率?

③ 粗产品中有哪些杂质? 如何除去?

④ 为减少溴乙烷的挥发损失，以提高产率，本实验采取了哪些措施?

3.3　2-氯丁烷的制备

3.3.1　目的

掌握2-氯丁烷的制备原理和操作方法。

3.3.2　原理

反应式

$$C_2H_5\underset{\large OH}{\underset{|}{C}}HCH_3 + HCl \xrightarrow{ZnCl_2} C_2H_5\underset{\large Cl}{\underset{|}{C}}HCH_3 + H_2O$$

3.3.3 实验仪器和主要试剂

（1）实验仪器 回流、分馏、蒸馏、萃取装置。

（2）主要试剂 $ZnCl_2$（无水），仲丁醇，浓盐酸，NaOH，$CaCl_2$（无水）。

3.3.4 实验步骤

① 在 100mL 圆底烧瓶上装好回流冷凝器及气体吸收装置。

② 向反应瓶中加入 16g 无水氯化锌和 7.5mL 浓盐酸，使其溶为均相[1]，冷却至室温。

③ 再加入 5mL 仲丁醇和 2 粒沸石，缓慢回流 40min。

④ 改用蒸馏装置，收集 115℃以下的馏分[2]。

⑤ 用分液漏斗分出有机相，依次用 6mL 水、2mL 5% NaOH 溶液、6mL 水洗涤。

⑥ 用无水氯化钙干燥。

⑦ 用分馏装置水浴加热，收集 67～69℃的馏分[3]。称量产品，计算产率。

纯 2-氯丁烷为无色透明液体，b. p. 68.25℃，d_4^{20} 0.8732，n_D^{20} 1.3971。

注释：

[1] 制备 Lucas 试剂时要先将无水氯化锌熔融、彻底干燥。因为无水氯化锌极易潮解，所以称量时应快速。

[2] 产物 2-氯丁烷不溶于酸，当反应瓶上层出现油珠状物质即为反应发生的标志。反应中生成的烯烃，在蒸馏时已从产物中除去。

[3] 2-氯丁烷沸点较低，操作时动作要快些，以免挥发而造成损失。

3.3.5 思考题

① 回流反应时为什么要缓慢回流？

② 为什么用分馏装置收集产品而不用蒸馏装置收集产品？

③ 实验中使产率降低的因素有哪些？

3.4 乙醚的制备

3.4.1 目的

学习分子间脱水法制备乙醚的原理和方法。

3.4.2 原理

在有机合成中，醚常作为重要的有机溶剂。通常有两种制备醚的方法：在浓酸的催化下发生分子间的脱水；及用卤代烃与碱金属盐的 Williamson 合成法。应注意：①脱水法主要用于制备具有低沸点的对称醚；②采用浓酸脱水催化法，可通过增加反应物浓度或减少产物使平衡右移。本实验采用浓酸催化达到使乙醇发生分子间脱水生成醚和水的目的，可以采用分水器分离生成的混合物中的水，反应式如下：

$$2CH_3CH_2OH \xrightleftharpoons[140℃]{H_2SO_4} CH_3CH_2OCH_2CH_3 + H_2O$$

纯乙醚是无色液体，沸点为 34.5℃，折射率为 1.3526，乙醚微溶于水，易溶于有机溶剂，极易挥发。

3.4.3 实验仪器和主要试剂

（1）实验仪器 三口瓶（150mL），滴液漏斗（150mL），圆底烧瓶（150mL），蒸馏头，温度计（200℃），减压蒸馏接收器，冷凝管，分液漏斗，量筒（25mL）。

（2）主要试剂 乙醇（95%），H_2SO_4（18.4mol/L），NaOH（1mol/L），饱和 NaCl 溶液，饱和 $CaCl_2$ 溶液，无水 $CaCl_2$。

3.4.4 实验步骤

3.4.4.1 制备

① 在干燥的150mL三口瓶中加入25mL乙醇（95%）和25mL浓 H_2SO_4。充分振动使混合均匀并加入几粒沸石。

② 如图3.1安装装置，三口瓶的一孔放上温度计，中间孔安上装有50mL乙醇的分液漏斗，侧口安装蒸馏头及冷凝管及接收装置，接液管支管连有橡皮管并通入下水道，检查反应装置是否严密。

③ 在石棉网上小心加热烧瓶中的反应液。当温度升高到140℃时开始滴加乙醇，控制滴加速度与馏出速度大致相等（约1滴/s），并维持反应温度在135～145℃之间。待乙醇加完后继续加热5～10min直到无馏分馏出。

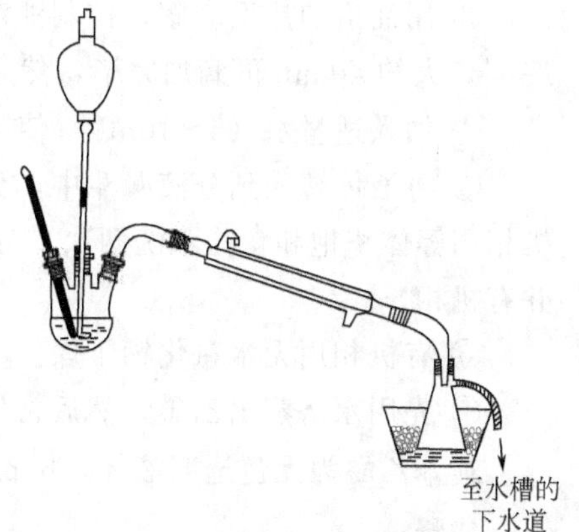

图3.1 制备乙醚的装置

3.4.4.2 精制

① 将接收瓶中的馏出液倒入分液漏斗中，先用等体积的NaOH（1mol/L）洗涤，静置分层；再用15mL饱和NaCl溶液洗涤，放出下层NaCl溶液层；最后用等体积的饱和 $CaCl_2$ 溶液洗涤两次，充分振荡，小心放出 $CaCl_2$ 溶液层。从分液漏斗上口将乙醚倾倒到小锥形瓶中，加2～3g颗粒状的无水 $CaCl_2$，并用玻璃塞塞紧，至溶液澄清。

② 将干燥后的乙醚在约50～60℃水浴中加热蒸馏，收集33～38℃的馏分。

3.4.5 思考题

① 如何准确地控制反应温度？如何确定反应接近终点？

② 进行每步洗涤的目的是什么？

3.5 苯乙醚的制备

3.5.1 目的

掌握制备苯乙醚的原理及操作方法。

3.5.2 原理

本实验是利用威廉姆逊（Williamson Reaction）反应合成苯乙醚。

反应式

$$PhOH + NaOH \longrightarrow PhONa + H_2O$$

$$PhONa + CH_3CH_2Br \longrightarrow PhOCH_2CH_3 + NaBr$$

3.5.3 实验仪器和主要试剂

（1）实验仪器 三口瓶，电动搅拌器，恒压滴液漏斗，回流、萃取、蒸馏装置。

（2）主要试剂 苯酚，氢氧化钠，溴乙烷，乙醚，无水氯化钙，饱和食盐水。

3.5.4 实验步骤

① 在 100mL 三口瓶上，装上电动搅拌器、回流冷凝管和恒压滴液漏斗。

② 将 3.75g 苯酚、2g 氢氧化钠和 2mL 水加入瓶中，开动搅拌器，使固体全部溶解。

③ 用油浴加热反应瓶，控制油浴温度在 80～90℃之间，并开始慢慢滴加 4.25mL 溴乙烷[1]，大约 40min 可滴加完毕，然后继续保温搅拌 1h，并降至室温。

④ 加入适量水（5～10mL）使固体全部溶解。

⑤ 将液体转入到分液漏斗中，分出水相，水相用 4mL 乙醚萃取一次，合并有机相。有机相用等体积饱和食盐水洗两次，分出有机相，并将水相每次都用 3mL 乙醚萃取一次，合并有机相。

⑥ 有机相用无水氯化钙干燥。

⑦ 先用水浴蒸出乙醚，然后常压蒸馏，收集 168～171℃的馏分[2]。

纯苯乙醚为无色透明液体，b.p. 170℃，d_4^{15} 0.9666，n_D^{20} 1.5076。

注释：

[1] 溴乙烷沸点低，实验时回流冷却水流量要大，以便保证有足够量的溴乙烷参与反应。

[2] 蒸去乙醚时不能用明火加热，将尾气通入下水道，以防止乙醚蒸气外漏引起着火。

3.5.5 思考题

① 制备苯乙醚时，萃取用饱和食盐水洗涤的目的是什么？

② 反应过程中，回流的液体是什么？出现的固体又是什么？为什么恒温到后期回流就不明显了？

3.6 苯甲醇和苯甲酸的制备

3.6.1 目的

① 学习利用 Cannizzaro 反应，用苯甲醛制备苯甲醇和苯甲酸的原理和方法。

② 掌握蒸馏和重结晶的操作技术。

3.6.2 原理

无 α-H 的醛与浓碱反应时可发生分子间的自身氧化还原，反应时一分子被氧化生成酸，另一分子被还原生成醇，这种反应称为 Cannizzaro 反应。

$$\text{C}_6\text{H}_5\text{CHO} + \text{NaOH} \longrightarrow \text{C}_6\text{H}_5\text{COONa} + \text{C}_6\text{H}_5\text{CH}_2\text{OH}$$

$$\text{C}_6\text{H}_5\text{COONa} \xrightarrow{H^+} \text{C}_6\text{H}_5\text{COOH}$$

3.6.3　实验仪器和主要试剂

（1）实验仪器　锥形瓶，烧杯，分液漏斗，圆底烧瓶，冷凝管，温度计，热过滤装置。

（2）主要试剂　苯甲醛，NaOH，HCl，无水 $MgSO_4$，饱和 $NaHSO_3$ 溶液，Na_2CO_3，乙醚。

3.6.4　实验步骤

① 在 100mL 烧杯中加入 8.8g NaOH 和 8.8mL H_2O，充分振荡，使其完全溶解[1]。

② 将 10.3mL 新蒸过的苯甲醛分批加入，充分振摇，使其成蜡状[2]。

③ 放置过夜。

④ 加入 30～50mL 水充分振摇。

⑤ 如不溶，补加少量水或温热使其全部溶解。

⑥ 冷却后混合物倒入分液漏斗中，分别用 24mL 乙醚分 3 次、每次 8mL 萃取，水相备用。合并乙醚相，分别用 4mL 饱和 $NaHSO_3$ 溶液、8mL 10% Na_2CO_3 溶液和 8mL 冷水洗涤[3]。

⑦ 提取液用无水 $MgSO_4$ 干燥。

⑧ 水浴蒸去乙醚[4]，再蒸馏收集 198～204℃馏分，即为苯甲醇。

⑨ 称重，计算产率。纯苯甲醇是无色液体，b. p. 205.3℃，d_D^{20} 1.0419，n_D^{20} 1.5396。

⑩ 在搅拌下将前面分出的水相以细流倒入 32mL 浓 HCl、32mL 水和 20g 冰的混合物中。

⑪ 冰水浴中冷却析出苯甲酸，减压收集苯甲酸。

⑫ 用水重结晶苯甲酸，干燥后测熔点。

⑬ 称重，计算产率。纯苯甲酸是白色针状晶体，m. p. 122.13℃。

注释：

[1] 使用浓碱时，操作要小心，尽量不要沾到皮肤上，若沾上要及时冲洗。

[2] 苯甲醛必须是新蒸馏过的，且分批加入。

[3] 产物中加入饱和 $NaHSO_3$ 溶液的目的是除去苯甲醛。

[4] 乙醚易燃，应远离火源。

3.6.5　思考题

① 为什么 Cannizzaro 反应在稀碱溶液中的反应比浓碱溶液中慢？

② 为什么饱和 $NaHSO_3$ 溶液能除去未反应的苯甲醛？

3.7　苯乙酮的制备

3.7.1　目的

① 学习用 Friedel-Crafts 酰化反应制备苯乙酮的原理和方法。

② 掌握蒸馏高沸点有机物的提纯技术

3.7.2 原理

Friedel-Crafts 酰化反应是制备苯乙酮的重要方法。苯与酰氯或酸酐作用是很好的合成反应。在本实验中，制备苯乙酮（苯基甲基酮）时，采用乙酐作酰化试剂。它比乙酰氯活性更好。

苯必须用无水 $CaCl_2$ 干燥，乙酐的沸点在 137～140℃。苯既作为反应物又和溶剂起作用，所以它是过量的。$AlCl_3$ 是催化剂，是 Lewis 酸，它可与产物形成稳定的复合物，其用量也要大于化学计量。

$$C_6H_6 + (CH_3CO)_2O \xrightarrow{AlCl_3} C_6H_5COCH_3 + CH_3COOH$$

3.7.3 实验仪器和主要试剂

（1）实验仪器　机械搅拌器，恒压滴液漏斗，冷凝管，真空泵，三口瓶，温度计，三角瓶，分液漏斗。

（2）主要试剂　苯，乙酸酐，无水 $AlCl_3$，HCl。

3.7.4 实验步骤

① 在 100mL 三口圆底烧瓶上，分别装上恒压滴液漏斗、电动搅拌装置和回流冷凝管[1]，冷凝管上端通过一氯化钙干燥管用橡皮管和一气体吸收装置相连（图 3.2）。

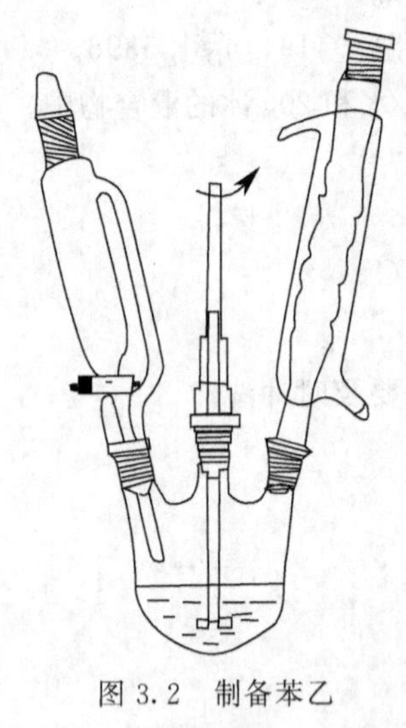

图 3.2　制备苯乙酮的装置

② 取下滴液漏斗，迅速往三口瓶中加入无水三氯化铝 10g[2] 和无水苯 16mL[3]，尽快塞好滴液漏斗。

③ 将乙酸酐 4mL[4] 加入滴液漏斗中，在搅拌条件下，逐滴加入三口烧瓶中。控制加入速度或用冷水冷却反应瓶（同时搅拌），勿使反应物剧烈沸腾（约需 10min）！加料完毕，反应剧烈程度稍减后，用 95℃左右的水浴加热反应瓶（同时搅拌），直至不再逸出 HCl 气体时为止（约需 50min）。

④ 待反应混合物冷至室温，在搅拌条件下，将反应物滴入盛有 18mL 浓盐酸[5] 和 30～40g 碎冰的烧杯中（在通风橱中进行），充分搅拌后若还有沉淀存在，再加适量浓盐酸使之溶解。

⑤ 分出上层有机层后，用乙醚萃取下层溶液两次，每次用量约 8～9mL。萃取液与上层液合并后，依次用 3mol/L 的 NaOH 8～9mL、水 9mL 洗涤，再用 2.5g 无水硫酸镁干燥。

⑥ 先在水浴上蒸馏回收苯和乙醚，稍冷后改用空气冷凝管在石棉网上加热，蒸馏收集 195～202℃的馏分，称重，计算产率。

苯乙酮为无色油状液体，b.p. 202℃，n_D^{20} 1.5372，产量 2.5～3.0g（产率 50%～60%）。

注释：

[1] 本实验所用的药品、仪器均应充分干燥（氯化氢吸收装置除外）。若要简化装置，可省掉电动搅拌器而采用二颈烧瓶进行反应。此时，可用电磁搅拌器进行搅拌或适当摇动反应装置而使反应顺利进行。

［2］本实验使用的无水三氯化铝应该是呈小颗粒或粗粉状，暴露于湿空气中立刻冒烟，滴少许水于其上则嘶嘶作响。称取和加入三氯化铝时，均应操作迅速，取用三氯化铝后，应立即将原试剂瓶塞好。

［3］化学纯苯经无水氯化钙干燥过夜后才能使用。

［4］所用乙酐必须在临用前重新蒸馏，取137～140℃馏分使用。

［5］加酸使苯乙酮析出，其反应式为：

$$C_6H_5C(CH_3)=\ddot{O}:AlCl_3 \xrightarrow{H^+/H_2O} C_6H_5C(CH_3)=\ddot{O}: + AlCl_3$$

3.7.5 思考题

① 为什么HCl出口应该远离水面或被浸入水中？

② 完成反应和获得高产率的关键是什么？如果装置或反应物含有少量的水分会对反应产生什么影响？

3.8 二苯甲酮（酰基化法）的制备

3.8.1 目的

熟悉和掌握制备二苯甲酮（酰基化法）的原理及操作方法。

3.8.2 原理

二苯甲酮是无色带光泽晶体，具有甜味及玫瑰香味，故用作香料，它能赋予香精以甜的气息，用在许多香水和香皂香精中。还可用于合成有机颜料、杀虫剂等，在医药工业上可用于生产苯甲托品氢溴酸盐以及苯海拉明盐酸盐等。

二苯甲酮的合成方法有很多，既可以用苄氯作原料经烷基化、氧化等反应制得，也可以由苯作起始原料通过烷基化、水解等步骤制备。本实验采取由苯甲酰氯和苯进行酰基化反应一步法制取二苯甲酮。

反应式

$$C_6H_5COCl + C_6H_6 \xrightarrow{AlCl_3} C_6H_5COC_6H_5 + HCl$$

3.8.3 实验仪器和主要试剂

（1）实验仪器　三口瓶，电动搅拌器，Y形管，气体吸收装置，回流装置，恒压滴液漏斗，萃取装置，减压蒸馏装置。

（2）主要试剂　无水三氯化铝，无水苯，苯甲酰氯，5%氢氧化钠水溶液，浓盐酸，无水硫酸镁。

3.8.4 实验步骤

① 在干燥的100mL三口瓶上分别装上电动搅拌器、回流冷凝管和Y形管。

② 在Y形管的两口分别装上恒压滴液漏斗和温度计，在冷凝管的上口接无水氯化钙的

干燥管并与气体吸收装置连接，在烧杯中放入5%氢氧化钠溶液作为吸收剂，吸收反应中产生的HCl气体，出气口与液面距离1～2mm为宜，千万不要全部插入液体中，以防倒吸。

③ 向反应瓶中迅速加入3.75g无水三氯化铝和15mL干燥的苯，在室温下边搅拌边自恒压滴液漏斗口向三口瓶慢慢滴加3mL苯甲酰氯，注意控制滴速使反应温度在40℃为宜。

④ 瓶内混合物开始剧烈反应，并伴有HCl气体产生，反应液逐渐变为褐色，滴加完毕后，在60℃水浴上加热并搅拌，至不再有HCl气体逸出为止，约需1.5h。

⑤ 待三口瓶冷却后，在通风橱内将反应物慢慢倒入盛有25mL冰水的烧杯中，有沉淀物析出。

⑥ 在搅拌下滴入1～2mL浓盐酸，直至沉淀物完全分解。

⑦ 用分液漏斗分出有机相，以苯作萃取剂对水相萃取两次（10mL×2）。

⑧ 合并有机相，依次用10mL水、10mL 5%氢氧化钠水溶液对有机相进行洗涤，然后再用水洗涤2～3次（每次10mL），直至有机相呈中性。

⑨ 用无水硫酸镁干燥，蒸除溶剂，得粗产物。

⑩ 减压蒸馏，收集187～189℃/2.00Pa（15mmHg）的馏分，得到无色透明液体，冰箱内冷却后固化，得到纯品。

粗产物也可用石油醚（60～90℃）重结晶代替减压蒸馏得到纯品。干燥后称量，测定熔点及红外光谱，计算产率。

二苯甲酮为无色晶体，m. p. 47～48℃，b. p. 305.4℃。

注释：

二苯甲酮有多种晶型，它们的熔点各不相同：α型为49℃，β型为26℃，γ型为45～48℃，δ型为51℃，其中α晶型较稳定。

3.8.5 思考题

① 在酰化反应中，是否容易产生多酰基取代芳烃？

② 与脂肪酮相比，芳酮分子中的羰基红外吸收峰是向高波数移动还是向低波数移动？为什么？

③ 为什么硝基苯可以作为傅-克反应溶剂？芳环上有OH、OR、NH_2等基团存在时对傅-克反应不利甚至不发生反应，为什么？

④ 为什么在减压蒸馏时，即使沸点较高的化合物最好也不用石棉网直接加热，而采用热浴加热？

3.9 环己酮的制备

3.9.1 目的

掌握氧化法制备环己酮的原理和方法。

3.9.2 原理

反应式

$$\text{C}_6\text{H}_{11}\text{—OH} \xrightarrow[H_2SO_4]{Na_2Cr_2O_7} \text{C}_6\text{H}_{10}\text{=O}$$

3.9.3　实验仪器和主要试剂

（1）实验仪器　回流、蒸馏、萃取装置。

（2）主要试剂　环己醇，重铬酸钠，氯化钠，浓硫酸，无水硫酸镁。

3.9.4　实验步骤

① 在100mL烧杯中加入30mL水和5.25g重铬酸钠，搅拌使之溶解，然后在冷却和搅拌下慢慢加入4.3mL浓硫酸，冷至30℃以下备用。

② 在100mL两口瓶中加入5.00g环己醇，然后将上述已溶解的重铬酸盐溶液加入其中，振荡使之混合。

③ 观察温度变化，当温度上升至55℃时，立即用冷水浴控制反应温度在55～60℃。大约0.5h后，温度开始下降，撤去冷水浴，将反应瓶放置1h，其间不断振荡，反应溶液变成墨绿色。

④ 在反应瓶中加入30mL水及沸石，安装成蒸馏装置[1]。

⑤ 将环己酮与水一起蒸出，环己酮与水能形成沸点为95℃的共沸混合物。

⑥ 直至馏出液不再浑浊，收集约25mL馏出液[2]。

⑦ 向馏出液中加入氯化钠使溶液饱和后，分出有机相。

⑧ 水相用30mL乙醚分两次萃取，萃取液与有机相合并，用无水硫酸镁干燥。

⑨ 在水浴上蒸去乙醚后，改用空气冷凝管进行常压蒸馏，收集151～155℃的馏分，产品3.0～3.50g，产率61%～71%。

纯环己酮为无色液体，b. p. 155.7℃，d_4^{20}0.9478，n_D^{20}1.4507。

注释：

[1] 加水蒸馏产品，实质上是一种简化了的水蒸气蒸馏。环己酮和水形成恒沸混合物，b. p. 95℃，含环己酮38.4%。

[2] 水的馏出量不宜过多，否则即使用盐析仍不可避免有少量环己酮溶于水中而损失。环己酮在水中的溶解度：31℃时为2.4g/100mL水。馏出液中加入食盐是为了降低环己酮在水中的溶解度，并有利于环己酮的分层。

3.9.5　思考题

① 重铬酸钠-浓硫酸混合物为什么冷至30℃以下使用？

② 盐析的作用是什么？

③ 该反应是否可以使用碱性高锰酸钾氧化？会得到什么产物？

3.10　己二酸的制备

3.10.1　目的

① 熟悉己二酸合成的原理。

② 掌握不同方法合成己二酸的基本操作。

3.10.2 原理

己二酸是合成尼龙-66 的主要原料之一，可以用高锰酸钾氧化环己酮制得，也可以用硝酸或高锰酸钾氧化环己醇制得。

反应式

方法 1 环己酮 $\xrightarrow{KMnO_4}$ $HOOC(CH_2)_4COOH$

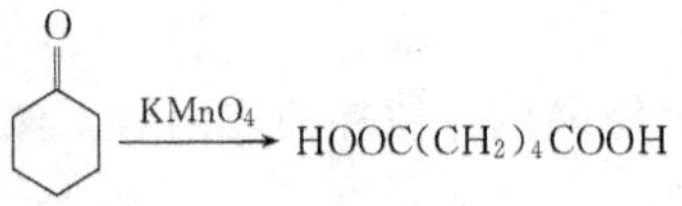

方法 2 环己醇 $\xrightarrow{[O]}$ 环己酮 $\xrightarrow{[O]}$ $HOOC(CH_2)_4COOH$

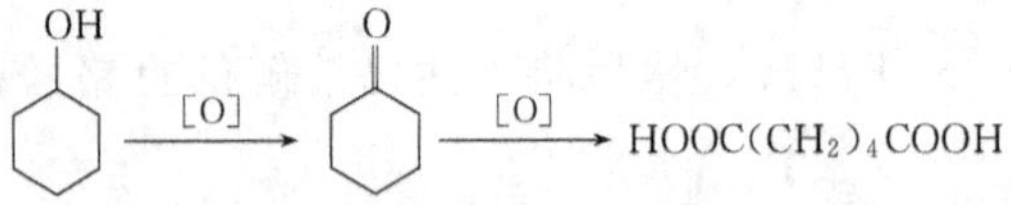

3.10.3 实验仪器和主要试剂

(1) 实验仪器　机械搅拌器，回流装置，抽滤装置，重结晶装置。

(2) 主要试剂

方法 1　环己酮，高锰酸钾，氢氧化钠，亚硫酸氢钠，浓盐酸。

方法 2　环己醇，高锰酸钾，碳酸钠，浓硫酸。

3.10.4 实验步骤

方法 1

① 在 100mL 三口瓶上分别安装搅拌器、温度计和回流冷凝管。

② 瓶内放入 6.3g 高锰酸钾、50mL 0.3mol/L 氢氧化钠溶液和 2mL 环己酮。注意反应温度，如反应温度超过 45℃时，应用冷水浴适当冷却，然后保持温度 45℃反应 25min，再在石棉网上加热至沸腾 5min 使反应完全。

③ 取 1 滴反应混合物放在滤纸上检查高锰酸钾是否还存在，若有未反应的高锰酸钾存在，会在棕色二氧化锰周围出现紫色环。假如有未反应的高锰酸钾存在则可加少量的固体亚硫酸氢钠直至点滴试验呈负性。

④ 抽气过滤反应混合物，用水充分洗涤滤饼。

⑤ 滤液置于烧杯中，在石棉网上加热浓缩到 10mL 左右。

⑥ 用浓盐酸酸化使溶液 pH=1～2 后再多加 2mL 浓盐酸，冷却后过滤，即得到粗产物。

⑦ 粗制的己二酸可用水重结晶，并用活性炭脱色，得到白色晶体 1.5g，产率 53%，熔点为 151～152℃。

纯己二酸为白色棱状晶体，熔点为 153℃。

方法 2

① 在 250mL 三口瓶上安装好搅拌器和温度计。

② 向反应瓶内加入 2.6mL 环己醇和碳酸钠水溶液（3.8g 碳酸钠溶于 35mL 温水）[1]。在搅拌下，分批加入研细的 12g 高锰酸钾，约需 2.5h。加入时控制反应温度始终大于 30℃[2]。

③ 加完后继续搅拌，直至反应温度不再上升为止。

④ 然后在 50℃水浴中加热并不断搅拌 30min。反应过程中有大量二氧化锰沉淀产生。

⑤ 将反应混合物抽滤，用10mL 10%碳酸钠溶液洗涤滤渣。

⑥ 搅拌下慢慢滴加浓硫酸，直到溶液呈强酸性，己二酸沉淀析出，冷却，抽滤，晾干。产量约2.2g，产率约60.2%，m.p.153℃。

注释：

方法1

此反应是放热反应，反应开始后会使混合物超过45℃。假如在室温下反应开始5min后，混合物温度尚不能上升至45℃，则可小心温热至40℃，使反应开始。

方法2

[1] 配制碳酸钠水溶液时水太少将影响搅拌效果，使高锰酸钾不能充分反应。

[2] 加入高锰酸钾后，反应可能不立即开始，可用水浴温热，当温度升到30℃时，必须立刻撤去温水浴，该放热反应自动进行。

3.10.5　思考题

① 为什么必须严格控制氧化反应的温度?

② 方法2的反应体系中加入碳酸钠有何作用?

3.11　苯甲酸的制备

3.11.1　目的

掌握苯甲酸的合成原理与基本操作。

3.11.2　原理

苯甲酸俗称安息香酸。苯甲酸及其钠盐是食品的重要防腐剂，苯甲酸可用作制药和染料的中间体，还用于制造增塑剂、聚酯聚合用引发剂、香料等，此外还可用作钢铁设备的防锈剂。

反应式

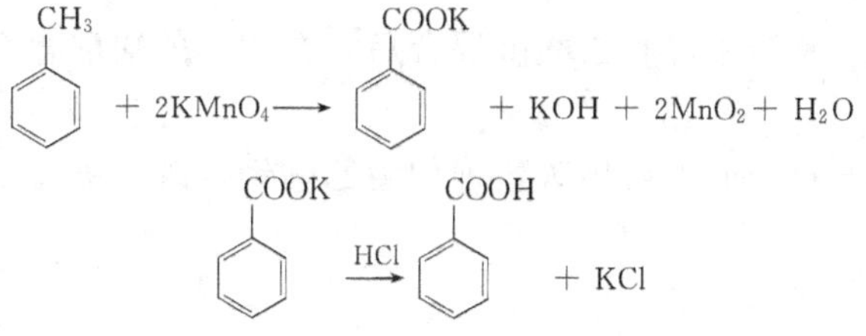

$$C_6H_5CH_3 + 2KMnO_4 \longrightarrow C_6H_5COOK + KOH + 2MnO_2 + H_2O$$

$$C_6H_5COOK \xrightarrow{HCl} C_6H_5COOH + KCl$$

3.11.3　实验仪器和主要试剂

(1) 实验仪器　回流、抽滤、重结晶装置。

(2) 主要试剂　甲苯，高锰酸钾，亚硫酸氢钠，刚果红试纸。

3.11.4　实验步骤

① 在250mL圆底烧瓶中加入2.7mL甲苯和100mL水，安装上回流冷凝管，在石棉网上加热至沸腾。

② 从冷凝管上口分批加入 8.5g 高锰酸钾，黏附在冷凝管内壁的高锰酸钾最后用 25mL 水冲洗入瓶内。

③ 继续加热煮沸并间歇振摇烧瓶，直到甲苯层几乎近于消失，回流液不再出现油珠（约需 4～5h）。

④ 将反应混合物趁热减压过滤[1]，用少量的热水（苯甲酸溶于热水，难溶于冷水）洗涤滤渣二氧化锰。

⑤ 合并滤液和洗涤液，放在冰水浴中冷却，然后用浓盐酸酸化至刚果红试纸变蓝，苯甲酸晶体析出。

⑥ 待溶液彻底冷却后，减压过滤出苯甲酸，用少量冷水洗涤，彻底抽干后，即得到粗产品[2]，干燥后产量约 1.7g。粗产品可在水中重结晶得到纯品。

纯苯甲酸为无色针状晶体，m.p.122.4℃。

注释：

[1] 滤液如果呈紫色，可加入少量亚硫酸氢钠使紫色褪去，重新减压过滤。

[2] 若苯甲酸的颜色不纯，可在适量的热水中重结晶，并用活性炭脱色。苯甲酸在 100g 水中的溶解度为：4℃ 0.18g；18℃ 0.27g；75℃ 2.2g。

3.11.5 思考题

① 还可以用什么方法来制备苯甲酸？

② 反应完毕后，若过滤液呈紫色，为什么要加入亚硫酸氢钠？

3.12 肉桂酸的制备

3.12.1 目的

① 学习通过 Perkin 反应制备肉桂酸的理论和方法。

② 学习回流和水蒸气蒸馏的基本操作。

3.12.2 原理

利用 Perkin 反应，将苯甲醛和乙酸酐混合后，在相应的羧酸盐存在下加热，可以制得 α，β-不饱和羧酸。

本实验进行 Perkin 反应时，可用碳酸钾代替乙酸钾，既可缩短反应时间还能提高反应产率。

$$C_6H_5CHO + (CH_3CO)_2O \xrightarrow[140\sim180℃]{K_2CO_3} C_6H_5CH{=}CHCOOH + CH_3COOH$$

3.12.3 实验仪器和主要试剂

(1) 实验仪器 圆底烧瓶，冷凝管，水蒸气蒸馏装置。

(2) 主要试剂 苯甲醛，乙酸酐，无水 K_2CO_3，HCl。

3.12.4 实验步骤

① 在 200mL 圆底烧瓶中放入 3mL 新蒸馏过的苯甲醛[1]、8mL 新蒸馏过的乙酸酐[2]以及研细的 4.2g 无水碳酸钾和几粒沸石，在石棉网上加热回流 30min[3]。

② 待反应物冷却后，加入 20mL 水，将瓶内生成的固体尽量捣碎（小心!），用水蒸气蒸馏蒸出未反应完的苯甲醛。

③ 再将烧瓶冷却，加入 10%氢氧化钠溶液 20mL，以保证所有的肉桂酸成钠盐而溶解。

④ 抽滤，将滤液倾入 250mL 烧杯中，冷却至室温，在搅拌下用浓盐酸酸化至刚果红试纸变蓝（pH 为 2～3）。

⑤ 冷却，抽滤，用少量水洗涤沉淀，抽干，自然晾干，产量约 3g（产率 65%～70%）。粗产品可用热水或 3∶1 的水-乙醇重结晶。纯肉桂酸（反式）为无色晶体，b.p.135.6℃ [4]。

注释：

[1] 苯甲醛放久了因自动氧化而生成较多量的苯甲酸，这不但影响反应的进行，而且苯甲酸混在产品中不易除干净，将影响产品的质量。故本反应所需的苯甲醛要事先蒸馏，截取 170～180℃馏分供使用。

[2] 乙酸酐放久了因吸潮和水解将转变为乙酸，故本实验所需的醋酐必须在实验前进行重新蒸馏。

[3] 由于有二氧化碳放出，反应初期有泡沫产生。

[4] 肉桂酸有顺反异构体，通常制得的是其反式异构体。

3.12.5 思考题

在无水丙酸钾的存在下，使用丙酸酐与苯甲醛反应会得到什么？

3.13 乙酸乙酯的制备

3.13.1 目的

① 通过乙酸乙酯的合成学会直接酯化反应。

② 进一步掌握蒸馏和萃取的操作技术。

3.13.2 原理

乙酸和乙醇在浓硫酸催化下直接酯化生成乙酸乙酯。反应是可逆的，通过加入过量的醇和除去产物如水来提高产率。

$$CH_3COOH + CH_3CH_2OH \xrightleftharpoons{H_2SO_4} CH_3COOCH_2CH_3 + H_2O$$

3.13.3 仪器和主要试剂

(1) 仪器　三口瓶（125mL），圆底烧瓶（50mL），冷凝管，锥形瓶，滴液漏斗（150mL），温度计（200℃），蒸馏头，接收管。

(2) 主要试剂　乙醇（95%），冰醋酸，H_2SO_4（18.4mol/L），饱和 NaCl 溶液，饱和 $CaCl_2$ 溶液，饱和 Na_2CO_3 溶液，无水 $MgSO_4$，石蕊试纸。

3.13.4 实验步骤

3.13.4.1 合成

① 在一干燥的125mL圆底烧瓶中放入6mL乙醇（95%），在冷却条件下，边摇动边缓慢滴加6mL H_2SO_4（18.4mol/L），混合均匀，投入几粒沸石。如图3.3安装仪器。漏斗末端和温度计水银球必须浸入液面以下，且距离瓶底0.5～1cm。将22mL乙醇（95%）和18mL冰醋酸混匀后，加入滴液漏斗。

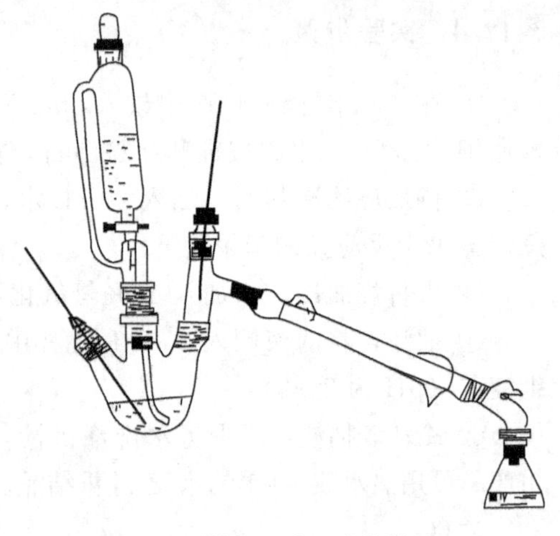

图3.3 制备乙酸乙酯的装置

② 用石棉网小火加热三口瓶，当反应温度升到110℃时，小心打开滴液漏斗，缓慢滴加乙醇和冰醋酸的混合物。控制滴液速度与馏出速度大致相等，并始终保持反应温度在110～125℃。滴加完毕，继续加热数分钟，直至温度升高到130℃，并不再有液体馏出为止。

3.13.4.2 精制

① 在馏出液中慢慢加入饱和碳酸钠溶液10mL（分批滴加）[1]，并不断振荡，直至不再有二氧化碳气体[2]（用石蕊试纸检查酯层，至不显酸性）。然后将混合液转入分液漏斗，静置5min，分去下层水溶液。

② 酯层用5mL饱和NaCl水洗涤[3]，分液后再用5mL饱和氯化钙溶液洗涤，分去下层液体。将酯层由上口倒入一干燥的锥形瓶中用3g无水硫酸镁干燥约15min。

③ 将干燥后的有机层液体进行蒸馏，加入沸石，收集73～78℃的馏分。称重并计算产率。

注释：

[1] 在馏出液的粗酯中含有一些乙酸，用饱和碳酸钠溶液除去。

[2] 可用检测溶液的pH至使石蕊试纸显蓝色的方法，来控制加入饱和碳酸钠溶液的量。

[3] 加入饱和NaCl水是为了帮助溶液更好地分层。

3.13.5 思考题

① 在本实验中，加入硫酸的目的是什么？

② 为了有利于酯化反应平衡，本实验采用了什么方法？

③ 粗酯中含有什么杂质？如何除去？

3.14 乙酸异戊酯的制备（香蕉油）

3.14.1 目的

掌握乙酸异戊酯的合成方法与原理。

3.14.2 原理

在人们的日常生活中，大部分酯具有广泛的用途。有些酯可作为食用油、脂肪、塑料以及油漆的溶剂。许多酯具有令人愉快的香味，是廉价的香料。

羧酸与醇或酚在无机或有机强酸催化下发生反应生成酯和水，这个过程称为酯化反应(Esterification Reaction)。常用的催化剂有浓硫酸、干燥的氯化氢、有机强酸或阳离子交换树脂。在酯化反应中，如果参与反应的羧酸本身就具有足够强的酸性，如甲酸、草酸等，就可以不另加催化剂。

该反应是一个可逆反应。在酯化反应中，如用等物质的量的有机酸和醇，反应达到平衡后，只能得到理论产量的67%。为了得到较高产量的酯，通常使用过量的酸或醇，促使平衡向产物方向移动。至于使用过量的酸还是过量的醇，取决于哪一种原料易得和价廉。另外，也可采用把反应中生成的酯或水及时地从体系中除去的方法来促使反应趋于完成。这可通过向反应体系中加入一些能与水形成低沸点共沸物的有机溶剂，如苯、甲苯、氯仿等，通过蒸馏共沸物带出生成的水。此外，还可用酰氯或酸酐与醇反应制取相应的酯。

反应式：

$$CH_3COOH + HOCH_2CH_2\underset{\substack{|\\CH_3}}{CH}CH_3 \xrightleftharpoons{H^+} CH_3COOCH_2CH_2\underset{\substack{|\\CH_3}}{CH}CH_3 + H_2O$$

3.14.3　实验仪器和主要试剂

(1) 实验仪器　回流、萃取、蒸馏装置，阿贝折光仪。

(2) 主要试剂　异戊醇，冰醋酸，浓硫酸，5%碳酸钠水溶液，饱和食盐水，无水硫酸镁。

3.14.4　实验步骤

① 将5mL异戊醇和7mL冰醋酸加入到25mL干燥的圆底烧瓶中，在振摇下缓缓加入1.0mL浓硫酸[1]，再投入几粒沸石，并配置回流冷凝管，加热回流1h。

② 回流结束后，冷却至室温，再将反应混合物从烧瓶中倒入分液漏斗。

③ 用10mL冷水洗涤反应瓶，洗涤液也倒入分液漏斗。振摇分液漏斗，静置分层，分出水层。

④ 有机层经5.0mL 5%碳酸钠水溶液洗涤后[2]，再用5.0mL饱和食盐水洗涤至中性[3]。

⑤ 分出水层后将酯层转入一个干燥的25mL锥形瓶中，用无水硫酸镁干燥。

⑥ 粗产物经过滤转入蒸馏瓶，蒸馏收集138～142℃馏分，接收瓶应插入冰浴中以减少气味。

⑦ 称量产品并计算产率，测定折射率。

乙酸异戊酯为无色透明液体，b.p.138～142℃，d_4^{20}0.876，n_D^{20}1.4000。

注释：

[1] 加入浓硫酸时，反应液会放热，应小心振荡，使热量迅速扩散。

[2] 碳酸钠水溶液洗涤粗产品时会产生二氧化碳气体，振摇时不宜剧烈，并留意放气。

[3] 一定要将有机相洗涤至中性，否则在蒸馏过程中产物易发生分解。

3.14.5　思考题

① 有利于乙酸异戊酯生成的一种方法是使冰醋酸过量，请提出另一种有利于酯生成平衡向右方向进行的方法。

② 过量的冰醋酸为什么比过量的异戊醇易从产物中除去？

3.15 乙酰水杨酸的制备（阿司匹林）

3.15.1 目的

① 学习乙酰化反应的基本原理和基本操作。

② 掌握重结晶的步骤。

3.15.2 原理

本实验是通过水杨酸的乙酰化反应制备乙酰水杨酸。乙酐作酰化剂，浓硫酸作催化剂，其作用是断裂水杨酸分子内氢键使乙酰化反应顺利进行。

主反应：

$$\text{C}_6\text{H}_4(\text{OH})(\text{COOH}) + (\text{CH}_3\text{CO})_2\text{O} \xrightleftharpoons{\text{H}^+} \text{C}_6\text{H}_4(\text{OCOCH}_3)(\text{COOH}) + \text{CH}_3\text{COOH}$$

粗产物含有的少量未反应的水杨酸可通过重结晶技术除去。

样品的纯度检验可用 $FeCl_3$ 溶液做对比实验确定。

乙酰水杨酸的含量可由 UV-751 紫外分光光度计测定。

3.15.3 实验仪器和主要试剂

（1）实验仪器　圆底烧瓶（150mL），温度计，冷凝管，烧杯（100mL、50mL），布氏漏斗，热水漏斗。

（2）主要试剂　水杨酸，乙酸酐，浓 H_2SO_4，乙醇（95%），$FeCl_3$（0.006mol/L）溶液。

3.15.4 实验步骤

3.15.4.1 制备

① 按图 2.14 安装装置制备乙酰水杨酸。

② 在干燥的 150mL 圆底烧瓶中加入 5.0g 干燥的水杨酸、10mL 新蒸的乙酸酐和 5 滴浓 H_2SO_4，慢慢摇匀，使水杨酸全部溶解。

③ 控制水浴温度在 70～80℃，加热 30min 后停止。

④ 稍冷，转移到 100mL 烧杯中，用 30mL 蒸馏水分几次冲洗烧瓶，合并于烧杯中，振荡，即有乙酰水杨酸结晶析出。如不析出结晶，可用玻璃棒摩擦瓶壁，并将反应物置于冰水中冷却使结晶完全。

⑤ 抽滤，用滤液反复淋洗锥形瓶，直至所有晶体被收集到布氏漏斗，用少量冷水洗涤结晶，继续抽滤将溶剂尽量抽干，制得粗阿司匹林。

3.15.4.2 精制

① 将结晶移到 100mL 烧杯中，加 6mL 乙醇（95%），水浴加热（60～70℃）至全溶，趁热过滤。

② 滤液中加入 15mL 水，冷却。

③ 充分冷却使结晶完全析出。

④ 抽滤并用玻璃塞压干晶体，再用少量冷乙醇洗涤 2 次。

⑤ 将结晶移到表面皿上，晾干。

⑥ 计算产率。

3.15.4.3　纯度检验

取几粒产品结晶加入盛有 5mL 水的试管中，加入 1～2 滴 $FeCl_3$（0.006mol/L）溶液，观察有无颜色反应。并与纯水杨酸作对比。

3.15.5　思考题

① 为什么制备乙酰水杨酸的装置必须是干燥的？

② 在乙酰化反应中为什么要加入浓硫酸？

③ 本实验中有什么副反应发生？

3.16　水杨酸甲酯的制备（冬青油）

3.16.1　目的

① 学习水杨酸甲酯的制备原理。

② 掌握水杨酸甲酯的制备操作技术。

3.16.2　原理

水杨酸甲酯是在 1843 年从冬青植物中提取出来的，故称冬青油，是一种天然酯，存在于依兰油、月下香油、丁香油中，具有冬青树叶的香气。常作香料，用于食品、牙膏、化妆品等。后来发现，水杨酸甲酯还具有止痛和退热的特性。水杨酸甲酯可由水杨酸和甲醇作原料在硫酸催化下酯化而得。

反应式

$$o\text{-}HOC_6H_4COOH + CH_3OH \xrightarrow{H_2SO_4} o\text{-}HOC_6H_4COOCH_3 + H_2O$$

3.16.3　实验仪器和主要试剂

（1）实验仪器　回流、蒸馏、减压、萃取装置，阿贝折光仪。

（2）主要试剂　水杨酸，甲醇，浓硫酸，10%碳酸氢钠。

3.16.4　实验步骤

① 在 30mL 干燥的圆底烧瓶中[1]，加入 3.45g 水杨酸和 15mL 甲醇。振摇溶解后，冷却下慢慢加入 1mL 浓硫酸，振摇反应瓶[2]，使反应物混合均匀，再加入 2 粒沸石。并配置回流冷凝管。

② 加热，回流 1.5h。

③ 然后将回流装置改为蒸馏装置，水浴加热，蒸除过量的甲醇。

④ 剩余反应混合物经冷却后加入 10mL 水，振摇后转入分液漏斗静置分层。

⑤ 分出有机层，水层用10mL乙醚萃取，合并有机相。

⑥ 有机相依次用10mL水、10mL 10%碳酸氢钠水溶液洗涤，然后再加水洗涤数次，使有机相呈中性。

⑦ 分出有机相，并用无水硫酸镁干燥。

⑧ 除去干燥剂，水浴蒸除乙醚[3]。将粗产物进行减压蒸馏，收集115～117℃/2.7kPa（20mmHg）的馏分。称量，测折射率并计算产率。

纯水杨酸甲酯为无色透明液体，m.p. −8～−7℃，b.p. 222℃，d_4^{25} 1.1787，n_D^{20} 1.5360。

注释：

[1] 反应容器需干燥，如含水会使酯化收率降低。

[2] 滴加浓硫酸时，如果没及时振摇反应瓶，有时会出现部分原料炭化现象。

[3] 要彻底蒸除甲醇，否则加水后产物溶解度增大，产率降低。

3.16.5　思考题

① 水杨酸与甲醇的酸催化酯化反应属于可逆反应，为了使反应正向进行，本实验采用了什么措施？还可以采取什么其他方法吗？

② 酯化反应结束后，如果不先蒸除甲醇而直接用水洗涤，将会对实验结果有何影响？

③ 在后处理中，为什么要用碳酸氢钠水溶液对粗产品进行洗涤？若用氢氧化钠水溶液来洗涤，将会产生怎样的情况？

④ 硫酸在此反应中的作用是什么？试写出水杨酸和甲醇的酸催化酯化反应机理。

3.17　乙酰乙酸乙酯的制备

3.17.1　目的

① 学习通过乙酸乙酯的Claisen缩合反应制备乙酰乙酸乙酯原理和方法。

② 掌握减压蒸馏的技术。

3.17.2　原理

从乙酸乙酯出发经Claisen缩合反应制备乙酰乙酸乙酯。

$$2CH_3COOC_2H_5 \xrightleftharpoons{C_2H_5ONa} CH_3\overset{O}{\overset{\|}{C}}CH_2\overset{O}{\overset{\|}{C}}OC_2H_5 + C_2H_5OH$$

3.17.3　实验仪器和主要试剂

（1）实验仪器　回流、减压蒸馏装置。

（2）主要试剂　Na，乙酸乙酯，乙酸，饱和NaCl溶液，Na_2CO_3，无水$MgSO_4$。

3.17.4　实验步骤

① 50mL干燥的圆底烧瓶中加入9.8mL无水乙酸乙酯[1]，然后迅速加入1.0g切细的金属钠[2,3]。

② 安装回流装置，水浴加热。在冷凝管上端安装一个氯化钙干燥管。

③ 加热反应混合物，控制反应温度使其慢慢回流至金属钠反应完（约需2h），此时溶

液应是呈透明红色。

④ 停止加热，使混合物冷却。

⑤ 在搅拌下，加 50%的乙酸[4]至反应液呈弱酸性（用石蕊试纸测试，pH=5～6），此时溶液中所有固体都已溶解。

⑥ 将反应液倒入分液漏斗中，加入等体积的饱和 NaCl 溶液，摇动静置。

⑦ 分出酯层，水层用 5mL 乙酸乙酯萃取，合并萃取液和酯层。

⑧ 用 5% 的 Na_2CO_3 溶液洗涤至中性。

⑨ 将有机层转移至锥形瓶中，用少量无水 $MgSO_4$ 干燥。

⑩ 将干燥后的液体轻轻倒出圆底烧瓶，在水浴下蒸馏，先蒸出乙酸乙酯。

⑪ 再减压蒸馏收集 54～55℃馏分/931Pa。

⑫ 称重，计算产率。

纯净的乙酰乙酸乙酯是无色液体，b. p. 180. 4℃（分解），d_D^{20} 1. 0282，n_D^{20} 1. 4194。

注释：

[1] 本实验所用的仪器需充分干燥，试剂应绝对无水。

[2] 除去金属钠表面的氧化物并将其切细，将加速缩合反应。

[3] 反应立即发生，若过于激烈，可用冷水浴冷却；若反应不立即开始，可直接加热促使反应发生，然后将热源移开。

[4] 加入乙酸时必须小心，以防止与未作用完的金属钠产生剧烈反应。

3. 17. 5　思考题

① 在这个实验中，为什么要求所使用的仪器和试剂必须是干燥的？

② 在制备乙酰乙酸乙酯的反应过程中，加入 50%的乙酸和饱和 NaCl 溶液的目的是什么？

3. 18　乙酰苯胺的制备

3. 18. 1　目的

① 熟悉通过乙酰化反应制备乙酰苯胺的方法和原理。

② 掌握有机物的回流和重结晶的基本技术。

3. 18. 2　原理

乙酰苯胺的制备可以在锌粉存在下，通过苯胺和乙酸发生乙酰化反应完成。

$$C_6H_5\text{—}NH_2 + CH_3COOH \rightleftharpoons C_6H_5\text{—}N(H)\text{—}C(=O)\text{—}CH_3 + H_2O$$

纯的乙酰苯胺是白色片状晶体（m. p. 114℃），难溶于冷水微溶于热水[1]，因此热水可作为提纯乙酰苯胺的溶剂。本实验使用的乙酸是过量的。

3. 18. 3　实验仪器和主要试剂

(1) 实验仪器　圆底烧瓶（150mL），冷凝管，Vigreux 分馏柱，安全瓶，温度计

(150℃)，烧杯（250mL)，剪子，吸滤瓶，布氏漏斗，真空泵，电热套，滤纸等。

(2) 主要试剂　苯胺（CP)，乙酸（CP)，锌粉，活性炭。

3.18.4　实验步骤

3.18.4.1　合成

① 按图 1.7 安装实验装置。

② 在 150mL 圆底烧瓶中加入 5mL 新蒸馏过的苯胺和约 0.1g 锌粉[2]，加热反应瓶约 1h 直至产生白雾为止，反应温度控制在 105℃。

③ 在不断搅拌下，将热反应物倒入含有 100mL 蒸馏水的烧杯中。冷却后，粗产品析出。析出完毕，将其减压过滤，用 5～10mL 冷水洗涤，洗净未反应的酸[3]。

3.18.4.2　精制

① 将粗产品用 100mL 热水溶解，加热至沸，如仍有油状物，则需补加热水直至溶解[4]。

② 停止加热，冷却几分钟，然后加入大约 1g 的活性炭[5]，在搅拌下加热脱色约 5min。将热的脱色后的溶液趁热过滤［如图 1.11 (a) 或 (b) 所示］。

③ 冷却滤液，乙酰苯胺逐步析出至不再有结晶析出。减压过滤收集结晶，用水洗涤两遍并用玻璃棒轻压。收集产品后放在表面皿上，自然晾干，称重，计算产率。

纯的乙酰苯胺的熔点是 114.3℃。

注释：

[1] 在 25℃，乙酰苯胺的溶解度是 0.0458mo/L，60℃是 0.0683mol/L，80℃为 0.2846mol/L，100℃为 0.4588mol/L。

[2] 加锌粉的目的是为了防止苯胺在反应过程中被氧化，反应中锌粉不要加过量，过量产生的不溶性的 $Zn(OH)_2$ 会影响下步反应的产率。

[3] 产物冷却后产品的固体结晶析出粘在反应器壁上，下一步反应很难处理。为避免出现这种现象，应在搅拌下迅速将热的反应物倒入冷的蒸馏水中，使过量的乙酸和未反应完的苯胺与之分离。

[4] 油珠不是杂质，是乙酰苯胺，因为当溶液温度超过 83℃时，少量的乙酰苯胺固体不溶于水中而以油状物存在。

[5] 在沸腾的溶液中加入活性炭，会引起暴沸，所以对热溶液应冷却几分钟后再加入。

3.18.5　思考题

① 在有机反应中，常用什么类型的化合物作乙酰化试剂？哪一个更为实用？

② 采用什么方法提纯固体有机物？如何用简单的方法确定物质的纯度？

3.19　对甲苯胺的合成

3.19.1　目的

掌握还原法制备对甲苯胺的原理与操作。

3.19.2　原理

利用芳香硝基化合物的性质，以金属为还原剂制备对甲苯胺。

反应式

$$CH_3-C_6H_4-NO_2 \xrightarrow[H^+]{Fe} CH_3-C_6H_4-NH_2$$

3.19.3 实验仪器与主要试剂

(1) 实验仪器 搅拌器，三口瓶，回流、蒸馏、萃取装置。

(2) 主要试剂 对硝基甲苯，还原铁粉，氯化铵，碳酸氢钠，苯，盐酸。

3.19.4 实验步骤

① 在50mL的三口瓶上安装搅拌器和回流冷凝管，向烧瓶中加入7.00g还原铁粉、0.90g氯化铵及20mL水[1]，边搅拌边加热，小火煮沸15min。

② 稍冷，加入4.50g对硝基甲苯，在搅拌下加热回流1h。

③ 反应结束后，冷却至室温，用5%碳酸氢钠溶液中和[2]。

④ 搅拌下将适量苯加入反应混合物内，抽滤，除去铁粉残渣[3]，用少量苯洗涤残渣。

⑤ 滤液倒入分液漏斗中，分出苯层，水相用苯萃取三次，合并苯萃取液。

⑥ 再用5%盐酸对上述苯萃取液提取三次，合并盐酸提取液。

⑦ 搅拌下往盐酸提取液中加入20%氢氧化钠溶液，析出粗产品。

⑧ 抽滤，并用少量水洗涤，再用少量苯萃取水相，苯萃取液与粗产品合并。

⑨ 水浴蒸馏除去苯，然后向残留物中加少量锌粉，在石棉网上加热蒸馏[4]，收集198～201℃馏分。产品约2.50g，m.p.44～45℃。

纯对甲苯胺为无色片状晶体，m.p.44～45℃，b.p.200.3℃，在空气及光的作用下因发生氧化作用而易变黑。

注释：

[1] 本实验以铁-盐酸作为还原剂，其中盐酸由氯化铵水解而得。$NH_4Cl + H_2O = NH_4OH + HCl$。

[2] 加入碳酸氢钠要控制pH在7～8之间，避免因碱性过强产生胶状氢氧化铁使分离发生困难。

[3] 铁残渣为活性铁泥，内含二价铁44.7%（以FeO计算），呈黑色颗粒状，在空气中会剧烈发热，故应及时倒入盛水的废物缸中。

[4] 除蒸馏法外，还可用乙醇-水混合溶剂重结晶法纯化对甲苯胺。

3.19.5 思考题

① 在还原反应开始前，为什么要对铁粉做预处理？

② 后处理时，为什么先加碳酸氢钠水溶液和苯，再用5%盐酸对苯层进行萃取？

3.20 邻氨基苯甲酸的合成

3.20.1 目的

掌握邻氨基苯甲酸的合成反应的原理和操作。

3.20.2 原理

反应式：

$$C_6H_4(CO)_2O + NH_3 \cdot H_2O \xrightarrow{\triangle} C_6H_4(CO)_2NH + 2H_2O$$

$$C_6H_4(CO)_2NH + Br_2 + 5NaOH \longrightarrow o\text{-}C_6H_4(COONa)(NH_2) + 2NaBr + Na_2CO_3 + 2H_2O$$

$$o\text{-}C_6H_4(COONa)(NH_2) \xrightarrow{CH_3COOH} o\text{-}C_6H_4(COOH)(NH_2) + CH_3COONa$$

反应机理：

$$C_6H_4(CO)_2NH + Br_2 + 5NaOH \longrightarrow o\text{-}C_6H_4(COONa)(NH_2) + 2NaBr + Na_2CO_3 + 2H_2O$$

$$C_6H_4(CO)_2NH \xrightarrow{NaOH} C_6H_4(COONa)(CONH_2) \xrightarrow{NaOBr/NaOH} [C_6H_4(COONa)(CONHBr)] \xrightarrow{-HBr} [C_6H_4(COONa)(CON:)] \longrightarrow [C_6H_4(COONa)(N{=}C{=}O)] \xrightarrow{H_2O} o\text{-}C_6H_4(COONa)(NH_2)$$

3.20.3 实验仪器和主要试剂

（1）实验仪器　回流装置，吸滤装置，重结晶装置。

（2）主要试剂　邻苯二甲酸酐，浓氨水，溴，氢氧化钠，浓盐酸，饱和亚硫酸氢钠溶液，冰醋酸；pH 试纸。

3.20.4 实验步骤

（1）邻苯二甲酰亚胺的制备　在 100mL 两口烧瓶中，放入 10g 邻苯二甲酸酐和 10mL 浓氨水，装上空气冷凝管及一支 360℃温度计。先在石棉网上加热，然后用小火直接加热，温度逐渐升到 300℃。间歇摇动烧瓶。用玻璃棒将升华进入冷凝管的固体物质推入烧瓶里。趁热把反应物倒入搪瓷盘中。冷却后凝成的固体放在研钵中研成粉末。产量约 8g，m.p. 232～234℃。

（2）邻氨基苯甲酸的制备　在 125mL 锥形瓶中，加入 7.5g 氢氧化钠和 30mL 水配制成的碱液。将此锥形瓶放入冰盐浴中，冷却至－5～0℃。往碱液中一次加入 2.1mL 溴[1]，振荡锥形瓶，使溴全部反应。此时温度略有升高。将制成的次溴酸钠冷却到 0℃以下，放置备用。

在另一个小锥形瓶中，加入 5.5g 氢氧化钠和 20mL 水配制的另一碱液。

取6g研细的邻苯二甲酰亚胺，加入少量水调成糊状物，一次全部加到冷的次溴酸钠溶液中，剧烈振荡锥形瓶。反应混合物应保持在0℃左右。从冰盐浴中取出锥形瓶，再剧烈摇动直到反应物转为黄色清液。把配制好的氢氧化钠溶液全部迅速加入，反应温度自行升高。把反应混合物加热到80℃约2min。加入2mL饱和亚硫酸氢钠溶液，以还原剩余的次溴酸。冷却，减压过滤。把滤液倒入250mL烧杯中，放在冰水浴中冷却。在不断搅拌下小心地滴加浓盐酸，使溶液恰好呈中性（pH=7，用石蕊试纸检验，约需15mL盐酸）[2]，然后再缓慢地滴加5～7mL冰醋酸，使邻氨基苯甲酸完全析出[3]。减压过滤，用少量冷水洗涤，晾干，产量约4g。灰白色粗产物用水进行重结晶，可得白色片状晶体。

纯邻氨基苯甲酸为白色片状晶体，m. p. 145℃。

注释：

[1] 溴具有强腐蚀性和刺激性，必须在通风橱中量取，取溴时应戴防护镜和橡皮手套，并且注意不要吸入溴蒸气。

[2] 邻氨基苯甲酸既能溶于碱，又能溶于酸，故过量的盐酸会使产物溶解。若加了过量的盐酸，需加氢氧化钠中和。

[3] 邻氨基苯甲酸的等电点约为3～4。为使邻氨基苯甲酸完全析出，必须加入适量的醋酸。

3.20.5　思考题

① 邻氨基苯甲酸在合成和分析上有哪些应用?

② 假若溴和氢氧化钠的用量不足或有较大的过量，对反应各有何影响?

③ 邻氨基苯甲酸的碱性溶液，加盐酸使之恰好呈中性，为什么不再加盐酸而是加适量醋酸使邻氨基苯甲酸完全析出?

3.21　苯氧乙酸的制备

3.21.1　目的

掌握苯氧乙酸的制备方法与原理。

3.21.2　原理

苯氧乙酸是一种白色片状或针状晶体，可用于合成染料、药物、杀虫剂，还可直接用作植物生长调节剂，且对人畜无害，因而应用较为广泛。苯氧乙酸可用苯酚、一氯乙酸在碱性溶液中进行威廉姆逊反应而制得。

总反应式

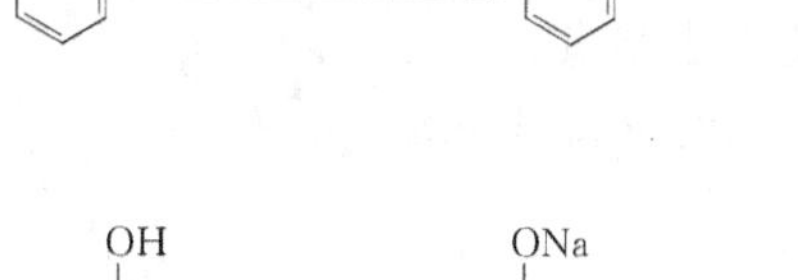

$$C_6H_5OH \xrightarrow{ClCH_2COOH/碱溶液} C_6H_5OCH_2COOH$$

分步反应式

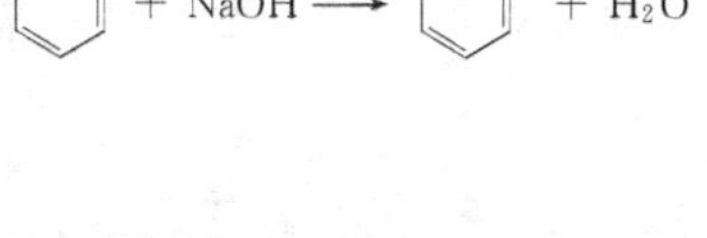

$$C_6H_5OH + NaOH \longrightarrow C_6H_5ONa + H_2O$$

$$2ClCH_2COOH + Na_2CO_3 \longrightarrow 2ClCH_2COONa + H_2O + CO_2$$

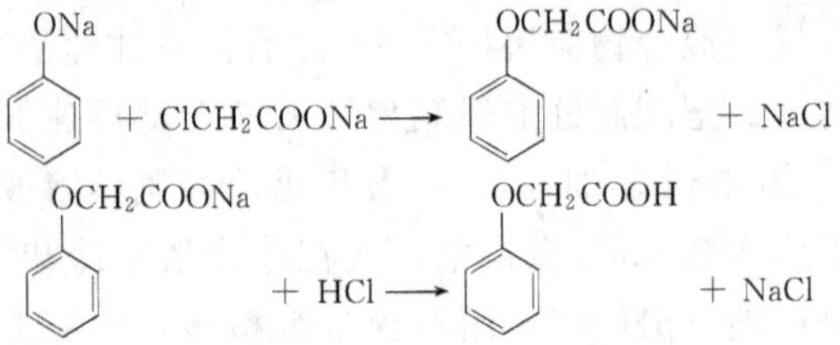

$$C_6H_5ONa + ClCH_2COONa \longrightarrow C_6H_5OCH_2COONa + NaCl$$

$$C_6H_5OCH_2COONa + HCl \longrightarrow C_6H_5OCH_2COOH + NaCl$$

3.21.3 实验仪器和主要试剂

（1）实验仪器　三口瓶，搅拌器，回流、萃取、抽滤装置。

（2）主要试剂　苯酚，氢氧化钠，一氯乙酸，碳酸钠，20%盐酸，乙醚，15%氯化钠溶液。

3.21.4 实验步骤

（1）配制氯乙酸钠溶液　依次将3.1g一氯乙酸[1]和10mL 15%氯化钠溶液加入到100mL烧杯中，搅拌下少量多次慢慢加入2g碳酸钠，加入速度以反应混合物温度不超过40℃为宜[2]。此时溶液pH为7～8，如不足此值，再改用饱和碳酸钠水溶液将反应混合液pH调至7～8。

（2）配制苯酚钠溶液　在100mL三口瓶上配置搅拌器[3]、回流冷凝管和温度计。向三口瓶中加入1.3g氢氧化钠、7.5mL水和2.8g苯酚[1]，开动搅拌器搅拌使固体溶解，冷却后待用。

（3）苯氧乙酸的合成

将配好的一氯乙酸钠溶液加入到上述苯酚钠溶液的三口瓶中，开动搅拌器，在石棉网上小火加热，使反应温度保持在100～110℃，回流2h[4]。

反应结束后，趁热将反应混合物倒入250mL烧杯中，加入30mL水，搅拌均匀，用浓盐酸调节溶液pH为1～2，冷却，析出白色晶体。抽滤，用5mL冷水洗涤粗产品，抽干后，将苯氧乙酸粗产品倒入250mL烧杯中，加入30mL水，加入固体碳酸钠使苯氧乙酸固体溶解。将溶液转入到分液漏斗中，加入10mL乙醚[5]，振荡、静置分层，除去乙醚层。水层再用20%盐酸酸化至pH为1～2，静置、冷却结晶，得到白色晶体，抽滤后用冷水洗涤滤饼两次，经干燥后即得精制产物。称量，测熔点，并计算产率。

纯苯氧乙酸为白色针状结晶，m.p.98～99℃。

注释：

[1] 一氯乙酸和苯酚具有腐蚀性，避免触及皮肤。

[2] 配制一氯乙酸钠溶液时，采用食盐水有利于抑制一氯乙酸钠的水解。中和反应温度超过40℃时，一氯乙酸易发生水解。

[3] 安装好搅拌器，避免损坏玻璃仪器。

[4] 合成苯氧乙酸反应刚开始时，反应混合物pH为12，随着反应的进行，其pH逐渐变小，直至pH为7～8，反应即告结束。

[5] 乙醚是用来萃取未反应而游离出来的少量酚。

3.21.5 思考题

① 以酚钠和一氯乙酸作原料制醚时，为什么要先使一氯乙酸成盐？可否用苯酚和一氯

乙酸直接反应制备醚？

② 用碳酸钠中和一氯乙酸时为何要加食盐水？

③ 在苯氧乙酸合成过程中，为何 pH 会发生变化，以 pH 为 7～8 作为反应终点的依据是什么？

④ 通过查阅文献，找出苯氧乙酸的其他合成方法。

3.22　甲基橙的制备

3.22.1　目的

① 学习利用苯磺酸和 N,N-二甲基苯胺，通过重氮盐的偶合反应制备甲基橙的方法和原理。

② 进一步掌握重结晶的技术。

3.22.2　原理

甲基橙常作为酸碱指示剂，本实验利用苯磺酸和 N,N-二甲基苯胺，通过重氮盐的偶合反应制备甲基橙。反应式如下：

$$H_2N-C_6H_4-SO_3H \xrightarrow{NaOH} H_2N-C_6H_4-SO_3Na \xrightarrow[0\sim5℃]{NaNO_2+HCl} HO_3S-C_6H_4-N_2Cl \xrightarrow[HAc]{C_6H_5-N(CH_3)_2}$$

$$HO_3S-C_6H_4-N{=}N-C_6H_4-N(CH_3)_2 \xrightarrow{NaOH} \underset{\text{甲基橙}}{NaO_3S-C_6H_4-N{=}N-C_6H_4-N(CH_3)_2}$$

3.22.3　实验仪器和主要试剂

（1）实验仪器　烧杯，热过滤漏斗，布氏漏斗等。

（2）主要试剂　对氨基苯磺酸，N,N-二甲基苯胺，亚硝酸钠，NaOH，HCl，冰醋酸，淀粉-碘化钾试纸。

3.22.4　实验步骤

3.22.4.1　重氮盐的制备

① 100mL 烧杯中放入 2.1g 对氨基苯磺酸晶体，加入 10mL 5%氢氧化钠溶液，玻棒搅拌下在热水浴中温热使之溶解[1]。

② 冷至室温后加入 0.8g 亚硝酸钠，使其溶解。

③ 搅拌下[2]，将上述混合溶液分批滴入盛有 13mL 水和 2.5mL 浓盐酸的烧杯中，并控制温度在 5℃以下[3]。

④ 滴加完后用淀粉-碘化钾试纸检验[4]。然后在冰盐浴中放置 15min，使重氮化反应完全。

3.22.4.2　偶合反应

① 取一支试管，加入 1.3mL N,N-二甲基苯胺和 1mL 冰醋酸，振荡使之混合。

② 不断搅拌下将此溶液慢慢加到上述冷却的重氮盐溶液中，加完后继续搅拌 15min，使偶合反应进行完全。

③ 搅拌下慢慢加入 15mL 10%氢氧化钠溶液，反应物变为橙色[5]，粗制的甲基橙呈细粒状沉淀析出。

④ 将反应物在沸水浴上加热 5min 使沉淀溶解，冷却至室温后再置于冰水浴中冷却，甲基橙全部重新结晶析出。

⑤ 抽滤，晶体用少量水洗涤，压干[6]。

⑥ 若要制得纯度较高的产品，可用溶有少量氢氧化钠（约 0.15g）的沸水重结晶。

⑦ 称重，计算产率。产量约 2g，产率约为 75%。

⑧ 溶解少许产品于水中，加几滴稀盐酸，然后用稀氢氧化钠溶液中和，观察溶液颜色有何变化。

纯甲基橙是橙黄色片状晶体，没有明确熔点。pH 3.1（红）～pH 4.4（橙黄）。

注释：

[1] 对氨基苯磺酸是两性化合物，其酸性略强于碱性，所以它能溶于碱中而不溶于酸中。

[2] 为了使对氨基苯磺酸完全重氮化，反应过程必须不断搅拌。

[3] 重氮化反应过程中控制温度很重要，若温度高于 5℃，则生成的重氮盐易水解成酚类，降低产率。

[4] 若不显蓝色，尚需酌情补加亚硝酸钠溶液。若亚硝酸已过量，可用尿素水溶液使其分解。

[5] 这时反应液呈弱碱性，若呈中性，则继续加入少量碱液至恰呈碱性，因强碱性又易生成树脂状聚合物而得不到所需产物。

[6] 湿的甲基橙在空气中受光照射后，颜色会很快变深，故一般得紫红色产物，如再依次以乙醇、乙醚洗涤晶体，可使其迅速干燥。

3.22.5 思考题

① 为什么本实验的实验过程的温度要低于 5℃？

② 二甲基苯胺与重氮盐发生偶合反应，为什么会形成对位产物？

③ 在重氮盐制备反应中，如果加入氯化铜会有什么结果？

3.23 苯佐卡因的制备

3.23.1 目的

① 掌握制备苯佐卡因的方法和原理。

② 了解多步骤合成的基本技术。

3.23.2 原理

苯佐卡因是对氨基苯甲酸乙酯的俗称，可作局麻药（Local Anesthetics）或止痛剂（Painkiller）。苯佐卡因的合成分两步完成，第一步是对硝基苯甲酸的还原，第二步是对氨基苯甲酸的酯化。

还原反应：

锡粉是还原剂，在酸性介质中，将对硝基苯甲酸还原成可溶于水的对氨基苯甲酸盐。生成的 $SnCl_4$ 通过加入浓氨水至碱性，生成沉淀被滤去。而生成的对氨基苯甲酸盐酸盐继续用冰醋酸中和，可析出对氨基苯甲酸固体。

$$p\text{-}O_2N\text{-}C_6H_4\text{-}COOH \xrightarrow[HCl]{Sn} p\text{-}HCl \cdot H_2N\text{-}C_6H_4\text{-}COOH + SnCl_4$$

$$SnCl_4 + 4NH_3 \cdot H_2O \longrightarrow Sn(OH)_4 \downarrow + 4NH_4Cl$$

$$p\text{-}HCl \cdot H_2N\text{-}C_6H_4\text{-}COOH \xrightarrow{NH_3 \cdot H_2O} p\text{-}H_2N\text{-}C_6H_4\text{-}COONH_4 \xrightarrow{CH_3COOH} p\text{-}H_2N\text{-}C_6H_4\text{-}COOH + CH_3COONH_4$$

酯化反应：

$$p\text{-}H_2N\text{-}C_6H_4\text{-}COOH \xrightarrow[H_2SO_4]{CH_3CH_2OH} p\text{-}H_2SO_4 \cdot H_2N\text{-}C_6H_4\text{-}COOC_2H_5 \xrightarrow{Na_2CO_3} p\text{-}H_2N\text{-}C_6H_4\text{-}COOC_2H_5$$

酯化产物与浓硫酸成盐而溶于水，反应完毕加碱中和即得苯佐卡因固体。

3.23.3　实验仪器和主要试剂

（1）实验仪器　三口瓶（150mL），圆底烧瓶（150mL），滴液漏斗（150mL），冷凝管，表面皿，烧杯，磁力搅拌器，布氏漏斗，电热套。

（2）主要试剂　对硝基苯甲酸，HCl，Sn，$NH_3 \cdot H_2O$，Na_2CO_3，H_2SO_4。

3.23.4　实验步骤

3.23.4.1　还原反应

① 在 150mL 三口烧瓶上安装回流冷凝管和滴液漏斗（图 3.4）。三口烧瓶中加入 4g 对硝基苯甲酸、9g 锡粉和磁力搅拌子，滴液漏斗加入 20mL（12mol/L）HCl。开动磁力搅拌器，从滴液漏斗中滴加 HCl，反应立即开始。如有必要可稍稍加热以维持反应正常进行（反应液中锡粉逐渐减少）。约 20～30min 后反应接近终点，反应液呈透明状。

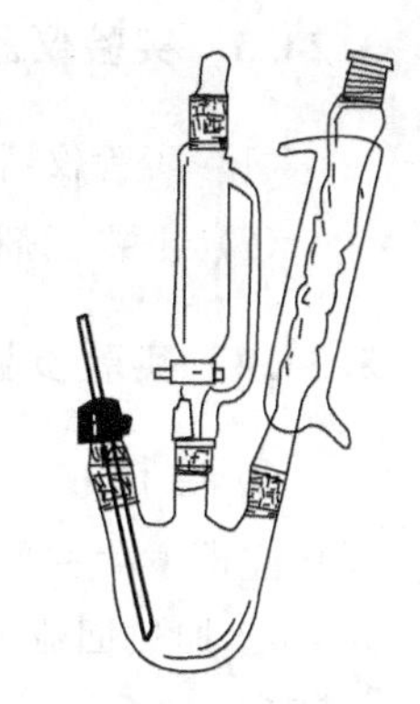

图 3.4　制备苯佐卡因的装置

② 稍冷后，将反应液倾入 250mL 烧杯中。待反应液冷至室温后，在不断搅拌下慢慢滴加 14mol/L 氨水，使溶液刚好呈碱性。用布氏漏斗抽滤以除去析出的氢氧化锡沉淀，用少许水洗涤沉淀，合并滤液和洗液，注意总体积不要超过 55mL，若体积超过 55mL，可加热浓缩。

③ 向滤液中小心地滴加冰醋酸，即有白色晶体析出。继续滴加少量冰醋酸，则有更多的固体析出，用蓝色石蕊试纸检验直到呈酸性为止。在冷水浴中冷却后抽滤得白色固体，晾干后称重，产量约为 2g。

纯对氨基苯甲酸为黄色晶体，m. p. 为 184～186℃。

3.23.4.2　酯化反应

① 在 150mL 三口烧瓶中加入 2g 对氨基苯甲酸、20mL 无水乙醇和 2mL 18.4mol/L H_2SO_4。将混合物充分摇匀，投入沸石，安上回流冷凝管，在电热套中加热回流 1h，反应液呈无色透明状。趁热将反应液倒入盛有 85mL 水的烧杯中。

② 溶液稍冷后，慢慢加入 Na_2CO_3 固体粉末，边加边用玻璃棒搅拌，使 Na_2CO_3 固体粉充分溶解。

③ 当液面有少许白色沉淀出现时，再慢慢滴加 1mol/L Na_2CO_3 溶液，将溶液 pH 值调至 9 左右。所得固体产品用布氏漏斗抽滤，晾干后称重。产量为 1～2g。

纯对氨基苯甲酸乙酯为白色针状晶体，m. p. 为 91～92℃。

3.23.5 思考题

① 如何判断还原反应达到终点？为什么？

② 试提出其他合成苯佐卡因的方法。讨论并比较。

3.24 对甲基-*N*-乙酰苯胺的合成

3.24.1 目的

掌握对甲基-*N*-乙酰苯胺合成的方法和原理。

3.24.2 原理

反应式：

$$H_3C-C_6H_4-NH_2 + (CH_3CO)_2O \longrightarrow H_3C-C_6H_4-NHCOCH_3$$

3.24.3 实验仪器和主要试剂

（1）实验仪器　回流、重结晶装置[1]。

（2）主要试剂　对甲苯胺，乙酸酐。

3.24.4 实验步骤

① 在 10mL 圆底烧瓶中，加入 2.00g 对甲苯胺和 2.4mL 乙酸酐[2]，装上回流冷凝管，反应立即发生并放热，使固体完全溶解。

② 加热回流 10min，趁热将反应混合物倒入 50mL 冷水中，边加边搅拌，立即析出淡黄色固体。

③ 彻底冷却后，过滤，用少量冷水洗涤晶体三次，干燥后得粗品 2.60g，产率 93%，m. p. 147～149℃。粗产品可用乙醇-水进行重结晶。

注释：

[1] 本实验所用仪器均是干燥的。

[2] 所用乙酸酐为新蒸馏的，收集 138～139℃的馏分。

3.25 安息香的合成

3.25.1 目的

掌握以苯甲醛为原料的多步合成反应的原理及操作。

3.25.2 原理

反应式

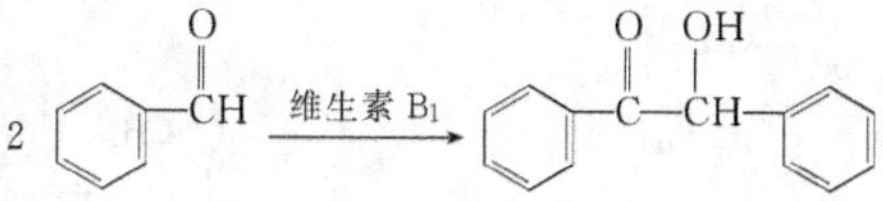

$$2\ C_6H_5CHO \xrightarrow{\text{维生素 } B_1} C_6H_5C(=O)CH(OH)C_6H_5$$

3.25.3 实验仪器和主要试剂

（1）实验仪器 重结晶、吸滤装置。

（2）主要试剂 苯甲醛，维生素 B_1，95%乙醇，氢氧化钠。

3.25.4 实验步骤

① 在 30mL 锥形瓶中，加入 0.30g 维生素 B_1 和 1.0mL 蒸馏水，待固体维生素 B_1 溶解后，再加入 3.0mL 95%乙醇。

② 塞上瓶塞，在冰盐浴中充分冷却[1]。

③ 将已冷却的 1.0mL 2.5mol/L 氢氧化钠溶液逐滴加入到上述冰盐浴的锥形瓶中，使溶液的 pH 为 10～11[2]。

④ 然后迅速加入 1.5mL 新蒸馏的苯甲醛，充分混匀，室温放置一天，有白色针状晶体析出。

⑤ 待结晶完全后，抽滤，用少量冷水洗涤晶体，晾干后得到粗产品。

⑥ 粗产品可用 95%乙醇重结晶。纯产品为白色针状晶体[3]，m.p.134～136℃。产量 0.6g，产率 38.5%。

注释：

[1] 维生素 B_1 市售品以其盐酸盐的形式储存。维生素 B_1 在酸性条件下稳定，但易吸水，受热易变质，在水溶液中易被空气氧化失效。遇光和 Cu、Fe、Mn 等金属离子均可加速氧化，所以应置于冰箱内保存。在 NaOH 溶液中噻唑环易开环失效，因此维生素 B_1 溶液、NaOH 溶液在反应前必须用冰水充分冷却，否则维生素 B_1 在碱性条件下会分解，这是实验成功的关键。没用完的维生素 B_1 应尽快密封保存在阴凉处。

维生素 B_1 是一种辅酶，可代替剧毒的氰化钠作催化剂进行安息香缩合，维生素 B_1 的结构

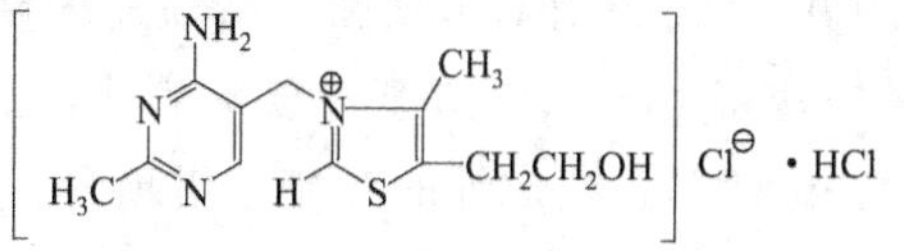

[维生素 B_1 结构式：NH_2，N，H_3C，N，CH_3，H，S，CH_2CH_2OH，$Cl^{\ominus}$ ·HCl]

维生素 B_1 分子中右边噻唑环上的氮原子和硫原子之间的氢有较大的酸性，在碱的作用下易被除去形成碳负离子，从而催化安息香的形成。本实验的反应机理如下。

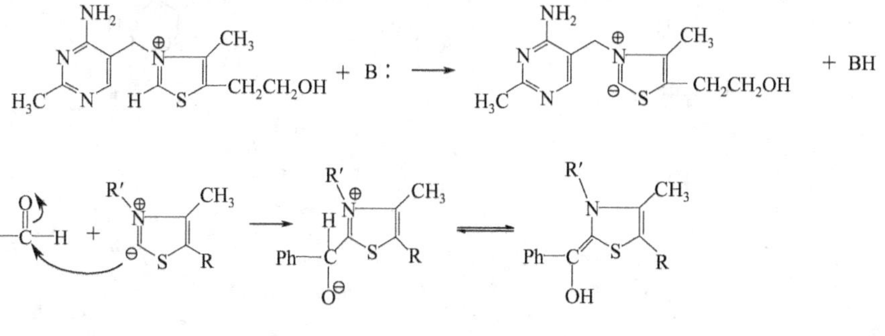

[反应机理式：维生素 B_1 + B: → 碳负离子 + BH；Ph—CHO + 噻唑碳负离子 → 加成物 ⇌ 烯醇式（Ph—C(OH)=）]

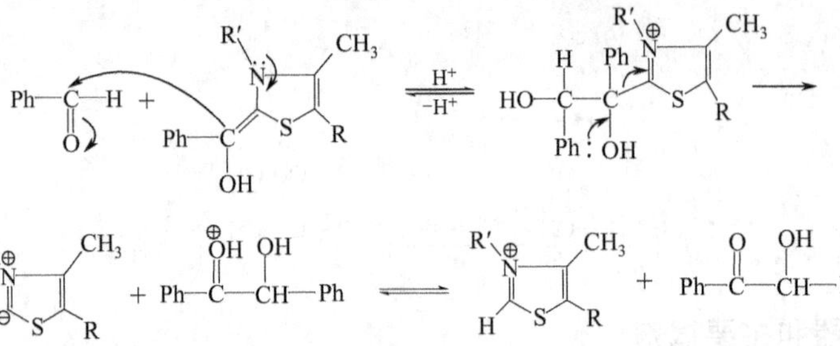

［2］控制 pH 是本实验的关键，因此应使用新蒸馏的苯甲醛。如苯甲醛中含有较多的苯甲酸，则影响溶液的 pH，可适当多加一些 NaOH 溶液，使 pH 保持在 10～11。

［3］安息香又称苯偶姻，是一种香料，DL 型为白色六边形单斜菱形结晶，D 型和 L 型都是针状结晶。

3.25.5 思考题

① 维生素 B_1 在安息香缩合反应中如何起催化作用？

② 氢氧化钠在此反应中起什么作用？

③ 试解析安息香的红外光谱图，并指出其主要特征吸收峰。

3.26 2-硝基雷琐酚的制备

3.26.1 目的

① 学习在苯环上进行亲电取代反应的定位规律及磺化反应的应用。

② 学习 2-硝基-1,3-苯二酚的合成方法。

3.26.2 原理

有机合成时，常引入其他基团，用以阻止或保护分子中某些潜在反应部位免受反应试剂进攻。这种基团必须导入容易，一些关键合成反应步骤完成后，又易于除去。

由雷琐酚（1,3-二羟基苯）合成 2-硝基雷琐酚，雷琐酚先磺化生成 4,6-二磺酸基雷琐酚，两个最易硝化部位被保护。将 4,6-二磺酸基雷琐酚硝化，然后水蒸气蒸馏二磺酸雷琐酚硝化物酸性溶液，水解除去磺酸基，生成纯的 2-硝基雷琐酚。

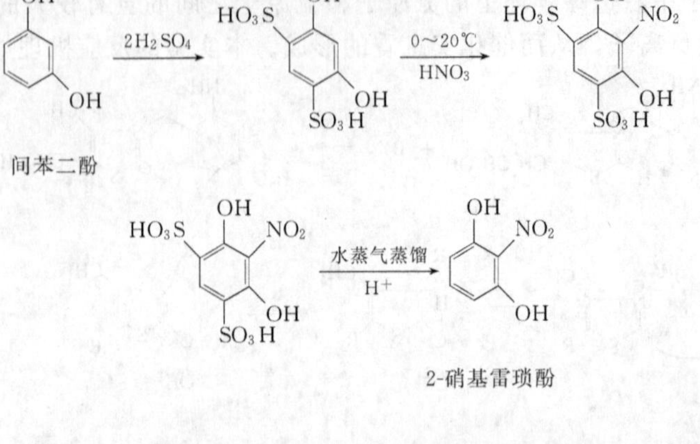

间苯二酚

2-硝基雷琐酚

3.26.3 实验仪器和主要试剂

（1）实验仪器 烧杯，滴液漏斗，水蒸气蒸馏装置，重结晶装置。

（2）主要试剂 粉状雷琐酚，浓硫酸，硝酸，浓硫酸，95%乙醇。

3.26.4 实验步骤

① 在150mL烧杯中，放置7.7g（0.07mol）粉状雷琐酚[1]，加28mL（50.4g，0.515mol）浓硫酸（98%，d=1.84）。几分钟后，如无黏稠的4,6-二磺酸雷琐酚浆状物生成，混合物加热到60～65℃。浆状物放置15min。

② 将4.4mL（4.38g，0.0693mol）硝酸（70%～72%，d=1.42）和6.2mL（11.9g，0.116mol）浓硫酸混合，混合酸用冰水浴冷却。

③ 浆状物用冰-盐浴冷至5～10℃，烧杯上方悬一滴液漏斗，搅拌浆状物，用滴液漏斗缓慢滴加已冷却好的混酸。控制滴加速度，使反应混合物温度不超过20℃。

④ 黄色混合物室温放置15min，然后保持混合物温度50℃以下，小心加入20g碎冰。

⑤ 将混合物转入500mL圆底烧瓶，加0.1g尿素[2]，进行水蒸气蒸馏。

⑥ 至冷凝管上无橘红色的2-硝基雷琐酚，或冷凝管上有不要的黄色针状4,6-二硝基雷琐酚（m.p.215℃）时，停止蒸馏。

水蒸气蒸馏5min左右，通常有产物出现。如蒸馏瓶中冷凝蒸气太多，产品难以蒸出。如果这样，停止通水蒸气，加热蒸馏瓶（煤气灯或加热套）。除去水，增加蒸馏瓶中的酸浓度，当瓶中酸的浓度足以催化脱磺酸基的反应时，停止加热，重新水蒸气蒸馏。

如冷凝管中充满固化产品；停止通冷凝水几分钟，直至产品熔化进入接收瓶。

⑦ 馏出液用冰水浴冷却，布氏漏斗过滤。

⑧ 稀乙醇重结晶。产品先溶于95%乙醇（3mL/g），趁热过滤，然后缓慢加水至溶液浑浊（少量沉淀），放置缓慢冷却。产量2.5～3.5g。

纯2-硝基雷锁酚的熔点84～85℃。

注释：

[1] 为磺化完全，雷锁酚应在研钵中研成很细的粉末。

[2] 过量硝酸与尿素成盐，溶于水而除去。

3.26.5 思考题

① 为什么磺化在4位和6位而不是2位？4,6-二磺酸基雷琐酚与雷琐酚比较，二磺化后活性（亲电取代反应）如何？

② 写出反应机理，解释脱磺酸基过程。

③ 设计一个机理，解释4，6-二雷琐酚的形成。

3.27 喹啉的制备

3.27.1 目的

学习用斯克劳普（Skraup）合成法制备喹啉的原理及操作方法。

3.27.2 原理

反应式

$$\underset{\mathrm{OH}}{\mathrm{CH_2}}-\underset{\mathrm{OH}}{\mathrm{CH}}-\underset{\mathrm{OH}}{\mathrm{CH_2}} \xrightarrow[-\mathrm{H_2O}]{\text{浓}\ \mathrm{H_2SO_4}} \underset{\mathrm{OH}}{\mathrm{CH_2}}-\mathrm{CH}=\underset{\mathrm{OH}}{\mathrm{CH}} \rightleftharpoons \underset{\mathrm{OH}}{\mathrm{CH_2}}-\mathrm{CH_2}-\underset{\mathrm{O}}{\overset{\|}{\mathrm{C}}}-\mathrm{H} \xrightarrow{\text{浓}\ \mathrm{H_2SO_4}} \mathrm{CH_2}=\mathrm{CH}-\underset{\mathrm{O}}{\overset{\|}{\mathrm{C}}}-\mathrm{H}$$

$$\mathrm{C_6H_5NH_2} + \mathrm{CHO{-}CH{=}CH_2} \longrightarrow \left[\mathrm{C_6H_5NH{-}CH_2{-}CH_2{-}CHO} \rightleftharpoons \mathrm{C_6H_5NH{-}CH_2{-}CH{=}CH{-}OH} \right]$$

$$\xrightarrow{-\mathrm{H_2O}} \text{1,2-二氢喹啉} \xrightarrow[\mathrm{C_6H_5NO_2}]{-\mathrm{H_2}} \text{喹啉}$$

喹啉可以通过苯胺、无水甘油、浓硫酸和弱氧化剂硝基苯等一起加热而制得，称斯克劳普（Skraup）合成法。为避免反应过于剧烈，常加入少量硫酸亚铁。

3.27.3 实验仪器和主要试剂

（1）实验仪器　回流、水蒸气蒸馏、萃取、蒸馏装置。

（2）主要试剂　无水甘油，苯胺，硝基苯，浓硫酸，30％氢氧化钠，硫酸亚铁，亚硝酸钠，乙醚，固体氢氧化钠，淀粉-碘化钾试纸。

3.27.4 实验步骤

① 在500mL圆底烧瓶中依次加入研成粉状的4g结晶硫酸亚铁[1]、29.9mL无水甘油[2]、9.3mL苯胺及6.7mL硝基苯，充分混合后，在振摇下缓缓加入18mL浓硫酸[3]。

② 装上回流冷凝管，用小火加热。当有小气泡产生并开始沸腾时，立即移去火源[4]。

③ 待反应趋于缓和时，再用小火加热，保持反应物和缓地沸腾2.5h。

④ 待反应液稍冷后，进行水蒸气蒸馏，除去未反应的硝基苯，直至馏出液不显浑浊为止。

⑤ 瓶中残留物稍冷后，加入30％氢氧化钠溶液，中和反应混合物中的硫酸，使溶液呈碱性[5]。

⑥ 再进行水蒸气蒸馏，蒸出喹啉及未作用的苯胺，直至馏出液变清为止。

⑦ 馏出液以浓硫酸酸化，待油状物全部溶解后，冰水浴中冷却至5℃左右。然后慢慢加入由3g亚硝酸钠和10mL水配成的溶液，直至取1滴反应液使淀粉-碘化钾试纸立即变蓝为止[6]。

⑧ 将混合物在沸水浴加热15min，至无气体放出。

⑨ 反应液冷却后，用30％氢氧化钠溶液碱化，再进行水蒸气蒸馏[7]。

⑩ 从馏出液中分出油层后，水层用乙醚萃取2次，每次用25mL乙醚。合并油层及乙醚萃取液，用氢氧化钠干燥过夜。

⑪ 蒸馏回收乙醚后直接加热蒸馏收集234～238℃的馏分[8]，产量8.10g[9]。文献值b.p.238℃，114℃/17mmHg。

注释：

[1] 硫酸亚铁的作用是防止反应物之间的迅速氧化，减缓反应的剧烈程度。

[2] 所用甘油的含水量不应超过0.50%（$d=1.26$），若甘油中含水量较大，则喹啉的产率不高。为除去甘油中的水分，可将普通甘油在通风橱内置于瓷蒸发皿中加热至180℃，冷至100℃左右，放入盛有硫酸的干燥器中备用。

[3] 试剂必须按所述次序加入，如果先加浓硫酸后加硫酸亚铁，则反应往往很剧烈，不易控制。

[4] 此反应系放热反应，反应液呈微沸表示反应已经开始，此时应停止加热。如果继续加热，则反应过于剧烈，会使溶液冲出容器。因此当反应太剧烈时，可用湿布敷于烧瓶上冷却。

[5] 每次碱化或酸化时，都必须将溶液稍加冷却，并充分搅拌后，再用试纸检验至呈明显的强碱或强酸性。

[6] 由于重氮化反应在接近完成时，反应进行很慢，故应在加入亚硝酸钠溶液2～3min后再检验是否有亚硝酸存在。

[7] 本实验系利用重氮化反应及重氮盐的特性，来除去喹啉中所夹杂的苯胺。此外，也可用对甲基苯磺酰氯去除粗制喹啉中的苯胺，或用氯化锌分离喹啉与未作用的苯胺。后一方法的原理是：在盐酸溶液中氯化锌虽与它们可形成复盐，但两者溶解度不同：

过滤收集析出的盐酸喹啉与氯化锌形成的复盐，加碱至最初形成的氢氧化锌沉淀复溶后，由醚提取喹啉即得。

[8] 最好在减压下蒸馏，收集110～114℃/14mmHg，118～120℃/20mmHg或130～132℃/40mmHg的馏分，得到无色透明的产品。

[9] 产率以苯胺计算，可不考虑硝基苯部分转化成苯胺而参与反应的量。

3.27.5 思考题

① 本实验中共使用3次水蒸气蒸馏操作，请回答下列问题：

a. 第一次水蒸气蒸馏的馏出液中，是否有苯胺和喹啉？为什么？

b. 第二次和第三次水蒸气蒸馏前为什么都要用碱中和反应液中的酸使溶液呈碱性？

c. 用什么简便的方法检验第二次水蒸气蒸馏的馏出液中是否含有苯胺？

② 在斯克劳普合成中，用对甲苯胺或邻甲苯胺代替苯胺作原料，各应得到什么产物？硝基化合物应如何选择？

③ 试说明本实验中影响产率和产品质量的主要因素。

3.28 尼可刹米的制备

3.28.1 目的

了解由羧酸和胺制备尼可刹米的原理及操作方法。

3.28.2 原理

反应式

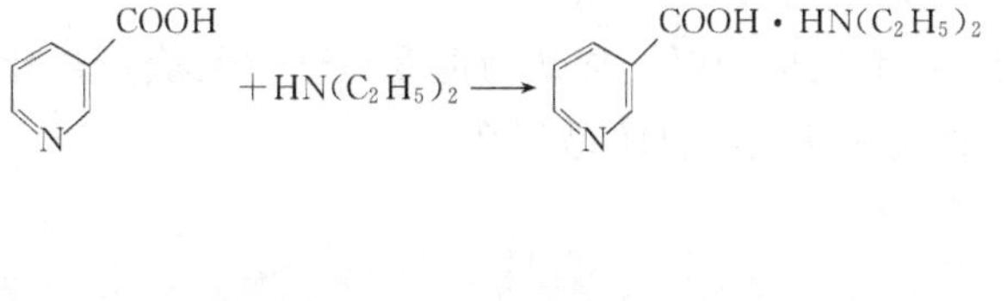

(3-吡啶甲酸，COOH) $+ HN(C_2H_5)_2 \longrightarrow$ (吡啶-3-位 $COOH \cdot HN(C_2H_5)_2$)

$$\text{(3-吡啶基)}COOH \cdot HN(C_2H_5)_2 \xrightarrow{POCl_3} \text{(3-吡啶基)}-CON(C_2H_5)_2 \cdot HCl + H_3PO_4$$

$$\text{(3-吡啶基)}-CON(C_2H_5)_2 \cdot HCl + NaOH \longrightarrow \text{(3-吡啶基)}-CON(C_2H_5)_2 + NaCl + H_2O$$

3.28.3 实验仪器和主要试剂

（1）实验仪器　三口瓶，搅拌器，分液漏斗，萃取装置，蒸馏装置，减压蒸馏装置。

（2）主要试剂　烟酸，二乙胺，三氯氧磷，10%高锰酸钾溶液。

3.28.4 实验步骤

① 在100mL干燥的三口瓶中，加入12.3g烟酸、10.2g二乙胺[1]，开动搅拌，慢慢加热，使固体物全部溶解[2]。

② 溶液冷至60℃以下，慢慢滴加8.4g三氯氧磷[3]，控制反应温度不超过140℃，滴完后维持135℃左右反应2.5h。

③ 将反应混合物冷至80℃，慢慢加入12mL水，待温度降至55℃后用20%氢氧化钠液中和[4]至pH 6～7，然后将反应液移至分液漏斗中，弃去水层。

④ 将油层移至100mL锥形瓶中，加10mL水稀释，再加入10%高锰酸钾溶液3mL，摇匀。将氧化后的反应液通过铺有活性炭（约3g）的漏斗脱色过滤。

⑤ 用少量水洗滤饼，洗涤液合并于滤液中，以10%的碳酸钾溶液调pH 7.5。将溶液转至分液漏斗中，用三氯甲烷提取4次（20mL×2，15mL×2），合并三氯甲烷层，用蒸馏水[5]洗涤4次（每次8mL）用无水碳酸钠干燥。

⑥ 将三氯甲烷提取液滤入50mL烧瓶中，先普通蒸馏除去三氯甲烷，再减压蒸馏收集160～170℃/10～15mmHg的馏分，得微黄液体12.5g。

文献值：m.p.24～26℃，b.p.175℃/25mmHg；158～159℃/10mmHg；128～129℃/3mmHg。

注释：

[1] 二乙胺及三氯氧磷用前要重蒸一次。烟酸应在80℃干燥过。

[2] 加料后如固体物已溶，则勿需加热

[3] 三氯氧磷易吸潮，放出氯化氢气体，故应在干燥条件下保存，宜在通风橱内蒸馏。

[4] 控制中和温度在60℃以下。

[5] 尼可刹米是药物，用自来水洗涤会引入其他杂质，影响产品的质量。

3.28.5 思考题

① 三氯氧磷在酰胺形成中起什么作用？

② 用氢氧化钠溶液中和反应液时，你认为温度高于60℃会产生什么结果？

③ 用10%高锰酸钾洗涤油层的目的是什么？

3.29　樟脑的还原反应

3.29.1　目的

① 学习用 $NaBH_4$ 还原樟脑的原理及操作方法。

② 了解薄层色谱在合成反应中的应用。

3.29.2　原理

用硼氢化钠还原樟脑得到冰片和异冰片 2 个非对映异构体。由于立体选择性较高，所得产物以异冰片为主。冰片和异冰片具有不同的物理性质，两者极性不同。

反应式

$NaBH_4$

樟脑　　冰片（龙脑）　+　异冰片

3.29.3　实验仪器和主要试剂

（1）实验仪器　锥形瓶，重结晶、蒸馏装置。

（2）主要试剂　樟脑，硼氢化钠，甲醇，乙醚，无水硫酸钠或无水硫酸镁。

3.29.4　实验步骤

（1）樟脑的还原　在 25mL 锥形瓶中将 1g 樟脑溶于 10mL 甲醇，室温下小心分批加入 0.6g 硼氢化钠[1]，一边振摇一边加硼氢化钠。必要时可用冰水浴控制反应温度。当所有硼氢化钠加完后，将反应混合物加热回流至硼氢化钠消失。将反应混合物冷却到室温，搅拌下将反应液倒在 20g 冰水中，待冰全部融化后，抽滤，收集白色固体，滤饼洗涤数次，晾干。将固体转移至 100mL 干净的锥形瓶中，加入 25mL 乙醚溶解固体，随后加入 6～7 刮刀无水硫酸钠或无水硫酸镁。干燥 5min 后将溶液（除去干燥剂）转移至预先称好的烧杯或锥形瓶中。在通风橱中蒸馏除去溶剂得到白色固体，并用无水乙醇重结晶。产量约为 0.6g，m.p. 212℃。

（2）产物的鉴别　取一片 5cm×15cm 的薄层板[2]，分别用冰片、异冰片、樟脑和樟脑还原产物的乙醚溶液点样，置于层析缸中展开[3]。取出层析板，待薄层上尚残留少许展开剂时，立即用另一块与薄层板同样大小并均匀地涂上浓硫酸的玻璃板覆盖在薄层板上，即可显色。将 4 个点的 R_f 值对比证明樟脑已被还原成冰片和异冰片。也可用溴化钾压片做产物的红外光谱。

注释：

[1] $NaBH_4$ 吸水后易变质，放出氢气，故开封后的试剂需置干燥器内保存。

[2] 薄层板的制法是：取 5g 硅胶与 13mL 0.5%～1%的羧甲基纤维素钠水溶液，在研钵中调匀，铺在清洁干燥的玻璃片上，薄层的厚度约 0.25mm。室温晾干后，在 110℃烘箱内活化半小时，取出放冷后置干燥器内备用。

[3] 以三氯甲烷-苯（2∶1，体积比）为展开剂。

3.29.5 思考题

① 测定产物熔点时应注意什么？

② 除薄层色谱外，还可用其他什么方法来区别和鉴别冰片和异冰片。

③ 原冰片酮用 $NaBH_4$ 还原时，你预计得到的主要产物是什么？

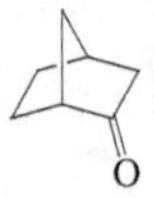

原冰片酮

3.30 降血脂药吉非贝齐中间体 3-(2,5-二甲基苯氧基)-1-氯丙烷的制备

3.30.1 目的

① 学习制备 3-(2,5-二甲基苯氧基)-1-氯丙烷的原理及操作方法。

② 了解薄层色谱在合成反应中的应用。

3.30.2 原理

在 NaOH 存在下，2,5-二甲基苯酚形成酚钠后与 1-氯-3-溴丙烷发生亲核取代反应，得 2,5-二甲基苯基-3-氯丙基醚。由于在反应条件下，反应物 2,5-二甲基苯酚钠与 1-氯-3-溴丙烷分别位于水相和有机相，故反应时间长、收率低。在相转移催化剂（PTC，如：TEBA）存在下，反应时间缩短，收率提高 30%以上。

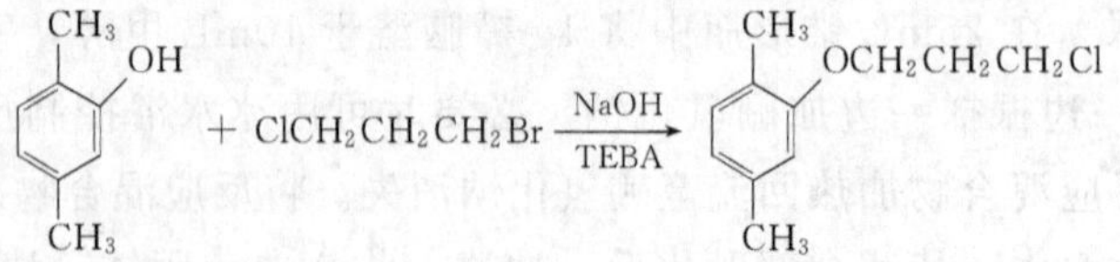

$$\text{2,5-}(CH_3)_2C_6H_3OH + ClCH_2CH_2CH_2Br \xrightarrow[\text{TEBA}]{\text{NaOH}} \text{2,5-}(CH_3)_2C_6H_3OCH_2CH_2CH_2Cl$$

3.30.3 实验仪器和主要试剂

（1）实验仪器　四口瓶，搅拌器，蒸馏、萃取、减压蒸馏装置。

（2）主要试剂　2,5-二甲基苯酚，1-氯-3-溴丙烷，NaOH，TEBA，乙醚，饱和氯化钠溶液，无水硫酸镁，苯，石油醚（60～90℃）。

3.30.4 实验步骤

① 在 100mL 四口瓶中加入 3.8g（0.031mol）2,5-二甲基苯酚、7.9g（0.05mol）1-氯-3-溴丙烷、0.2g TEBA，加热，维持 90～94℃快速搅拌下滴加 1.6mol/L NaOH[1]，约 1h 滴完（滴加 1.6mol/L NaOH 约 30mL）。滴完后维持 100℃快速搅拌至近中性（pH 6～7，约搅拌 4h）[2]，冷至室温，分液，水层用乙醚（15mL，10mL×2）萃取，合并有机层，用饱和氯化钠溶液洗涤，无水硫酸镁干燥。普通蒸馏回收乙醚后减压蒸馏，收集 108～110℃/266Pa 的馏分，得亮黄色液体 4.7g，收率 77.7%[3]。

② 薄层色谱分析展开条件

展开剂：苯-石油醚（1∶1）。

吸附剂：硅胶 GF254。

展开距离：10cm。

显色方法：紫外光。

注释：

[1] 因该反应为相转移催化反应，故应维持较快的搅拌速度。

[2] 如果 pH 偏小，应补滴加少量 NaOH。

[3] 在实验条件下，2,5-二甲基苯酚钠与 1-氯-3-溴丙烷反应除了得到产物 3-(2,5-二甲基苯氧基)-1-氯丙烷外，还有少量 3-(2,5-二甲基苯氧基)-1-溴丙烷生成，其沸点为 124～132℃/266Pa。

3.30.5 思考题

① 2,5-二甲基苯酚钠与 1-氯-3-溴丙烷反应为什么要加相转移催化剂？

② 写出下列醚的 Williamson 合成法。

A. $CH_3CH_2OCH_2CH_3$

B. $CH_3CH_2OCH(CH_3)_2$

C. $CH_3CH_2CH_2CH_2OCH_2CH_2CH_3$

③ 为什么只有少量 3-(2,5-二甲基苯氧基) -1-溴丙烷生成？

④ 为什么应维持较快的搅拌速度？

第二部分　提取、分离与鉴定

3.31 苯甲酸的提纯

3.31.1 目的

① 学习重结晶提纯固体有机化合物的原理和方法。

② 掌握重结晶的实验操作。

3.31.2 原理

重结晶（有关原理和操作详见 2.6.1）是提纯固体有机物常用的方法之一。

固体化合物在溶剂中的溶解度随温度变化而变化，一般温度升高溶解度增加，反之则溶解度降低。重结晶利用这一性质，将含有杂质的固体产品溶解在热溶剂中制成饱和溶液，通过热过滤将不溶性杂质除去，滤液冷却后由于溶解度降低，溶液就会变得过饱和而析出结晶，而杂质全部或大部分仍留在溶液中，通过过滤除去可溶性杂质，达到提纯的目的。

纯苯甲酸为无色针状晶体，熔点 122.4℃，沸点 249℃，100℃易升华，溶于乙醇、乙醚、氯仿和苯等有机溶剂，微溶于水。

本实验利用苯甲酸微溶于冷水、易溶于热水的性质，选择水作溶剂进行重结晶提纯。

3.31.3 实验仪器和主要试剂

(1) 实验仪器　圆底烧瓶（250mL），冷凝管，烧杯，热过滤漏斗，布氏漏斗，吸滤瓶等。

（2）主要试剂　苯甲酸，活性炭，蒸馏水。

3.31.4　实验步骤

① 按图 2.14 安装回流装置。

② 加 3.0g 不纯的苯甲酸于 250mL 圆底烧瓶中，加 80mL 蒸馏水然后加热至沸。待溶液稍稍冷却，加入大约 0.1g 活性炭重新加热沸腾几分钟。

③ 用热过滤方法趁热过滤溶液，皱形滤纸叠法见图 2.4。用烧杯收集滤液。

④ 将滤液放置并冷却使晶体完全析出。

⑤ 减压过滤分离苯甲酸晶体，用少量冷水洗涤（见图 2.2）。

⑥ 将晶体摊放在滤纸上使其自然干燥。

⑦ 称重并计算产率。

3.31.5　思考题

① 重结晶的过程分几个步骤？每步骤的目的是什么？

② 为什么重结晶使用的溶剂不能太多也不能太少？应该如何准确控制用量？

③ 为什么在溶解提纯物时溶剂的用量要少于计算量？

3.32　烟碱的提取与鉴定

3.32.1　目的

① 掌握水蒸气蒸馏等基本操作技术。

② 熟悉烟碱的提取原理，了解烟碱的性质。

3.32.2　原理

烟碱又名尼古丁，存在于烟叶中。结构式为：

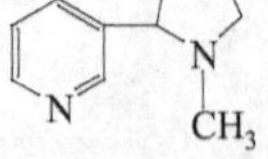

N　N　CH_3

烟碱在常温下为无色或淡黄色液体，沸点 247℃，有剧毒。烟碱有很强的碱性，易与酸作用生成盐溶于溶液中。本实验利用此性质，使烟碱与盐酸生成烟碱的盐酸盐溶于溶液中，中和至碱性，然后用水蒸气蒸馏提取分离得到烟碱的水溶液。

烟碱可与酚酞试剂作用，可被氧化剂氧化。烟碱同其他生物碱一样可与生物碱试剂发生沉淀反应。

3.32.3　实验仪器和主要试剂

（1）实验仪器　水蒸气蒸馏、吸滤装置，pH 试纸，试管，量筒（50mL），烧杯（250mL）。

（2）主要试剂　盐酸（3mol/L），氢氧化钠（6mol/L），酚酞，高锰酸钾（0.3mol/L），饱和苦味酸溶液，醋酸（8mol/L），碳酸钠（0.5mol/L），碘液，碘化汞钾。

3.32.4　实验步骤

3.32.4.1　烟碱的提取

用台秤称取2.0g烟丝于250mL烧杯中，加40mL盐酸（3mol/L），加热煮沸20min，不断搅拌，同时注意补充水以保持液面不下降。抽滤，滤液用6mol/L氢氧化钠中和至碱性（可用pH试纸测试）。

3.32.4.2　水蒸气蒸馏

将烟碱提取液进行水蒸气蒸馏（装置见图2.9，有关操作详见2.5.3），收集约30mL蒸馏液，可停止水蒸气蒸馏。

切记：停止蒸馏时，必须先旋开螺旋夹，然后移去热源！

3.32.4.3　烟碱的性质实验

（1）碱性　取一支试管加1mL烟碱水溶液，加1滴酚酞观察现象并解释（也可用pH试纸试验）。

（2）氧化反应　取一支试管加入5mL烟碱水溶液，再加入1滴0.3mol/L高锰酸钾溶液和3滴0.5mol/L碳酸钠溶液，振荡试管，注意观察溶液的颜色变化，有无沉淀生成？

（3）沉淀反应

① 取一支试管加入5mL烟碱水溶液，再加入5滴饱和溶液苦味酸，振荡试管，观察现象。

② 取一支试管加入5mL烟碱水溶液，再加入5滴碘液，振荡试管，观察现象。

③ 取一支试管加入5mL烟碱水溶液，再加入1滴HAc然后慢慢加入碘化汞钾溶液，振荡试管，观察现象。

3.32.5　思考题

① 简述水蒸气蒸馏的基本原理。

② T形管和安全管的作用是什么？

③ 如何判定水蒸气蒸馏已经完成？

3.33　从茶叶中提取咖啡因

3.33.1　目的

① 学习从茶叶中提取咖啡因的提取方法、主要原理和基本操作。

② 熟悉升华的过程。

3.33.2　原理

咖啡因的提取既可用微波萃取法也可用传统的索氏萃取法。

咖啡碱又名咖啡因，属杂环化合物嘌呤的衍生物，它的化学名称为1,3,7-三甲基-2,6-二氧嘌呤，结构如下：

嘌呤

咖啡因（1,3,7-三甲基-2,6-二氧嘌呤）

3.33.3 实验仪器和主要试剂

（1）实验仪器　微波炉，蒸馏装置，索氏提取装置，升华装置。

（2）主要试剂　茶叶，乙醇（95%），生石灰粉。

3.33.4 实验步骤

3.33.4.1 方法一：微波萃取法[1]

① 称取研细的茶叶末 10g，置于 250mL 碘量瓶中，加入 120mL 的乙醇（95%），放入沸石。

② 将碘量瓶放于普通微波炉中，调节功率约 320W，辐射约 50～60s[2]，取出冷却。

③ 重复上述步骤 3～4 次[3]，过滤，除去茶叶末。

④ 冷却后改用水浴蒸馏（见 2.5.1）蒸出提取液中的大部分乙醇（可回收利用），至提取液的残液为 5～8mL 时停止蒸馏。

⑤ 将残液倒入蒸发皿中，并用蒸出的乙醇对蒸馏烧瓶稍作洗涤，一并倒入蒸发皿中。

⑥ 加 2.5g 生石灰粉[4]，不断搅拌，并将蒸发皿置于水蒸气浴上蒸干溶剂。

⑦ 将蒸发皿移至石棉网上，小火加热，不断焙炒至干[5]。

⑧ 取一张稍大些的圆形滤纸，罩在大小适宜的玻璃漏斗上，刺上小孔，再盖在蒸发皿上，漏斗颈部塞入少许棉花。

⑨ 用小火慢慢加热升华[6]（见 2.6.2），当有棕色油状物在玻璃漏斗壁上生成时，立刻停止加热，冷却，收集滤纸上的咖啡碱晶体。

⑩ 残渣经充分搅拌后，用略大的火再升华 1～2 次，合并数次升华产物，称重，产量约 70～80mg。咖啡碱为白色或略带微黄色的针状晶体，m.p. 238℃。

3.33.4.2 方法二：索氏萃取法

① 称取茶叶末 10g，置于合适的滤纸筒中，然后放入索氏提取器内[7]。

② 取 60mL 乙醇（95%）于圆底烧瓶中，放入沸石，安装好回流提取装置（图 2.6），水浴加热。

③ 连续回流提取 2h 左右，直至提取液颜色较淡，当溶液刚好虹吸回流至烧瓶中时，即可停止加热[8]。

④ 其余步骤按方法一中的第③～⑨步操作[9]。

注释：

[1] 微波萃取法比其他方法所需的实验时间可缩短 2h 左右。

[2] 微波辐射时间以不使溶液暴沸冲出为原则。

[3] 重复微波辐射要先冷却。

[4] 生石灰起中和作用，以除去丹宁酸等酸性物质。

[5] 若残留少量水分，则会在下一步升华开始时漏斗壁上呈现水珠。如有此现象，则应撤去火源，迅速擦去水珠，然后继续升华。

[6] 升华操作直接影响到产物的质量与产量，升华的关键是控制温度。温度过高，将导致被烘物冒烟炭化，或产物变黄，造成损失。

[7] 滤纸筒的大小要紧贴器壁，高度不要超过索氏提取器的虹吸管。

[8] 控制回流速度，一般 2h 内虹吸 8～10 次。

[9] 产物约为 30～40mg。

3.33.5　思考题

① 假定你得到的是纯咖啡因。茶叶中含咖啡因约 3%，本实验你得到咖啡因的产量是多少？与理论值比较差距有多少？给出合适的解释。

② 对比微波法和 Soxhlet 提取法，并简述它们的特点。

3.34　橘皮中有效成分的提取、分离与鉴定

3.34.1　目的

① 掌握植物中有效成分的提取方法和原理。

② 进一步巩固水蒸气蒸馏、萃取等基本操作技术。

3.34.2　原理

柠檬、橙子与柑橘等水果的新鲜果皮中含有一种香精油，叫柠檬油（lemon oil）。果皮中含油量 0.35%。柠檬油为黄色液体，有浓郁的柠檬香气 $\rho=0.857\sim0.862$（15/4℃）。n_D^{20} 1.474～1.476. $[\alpha]_D^{20}=+57°\sim+61°$。主要成分是柠烯，含量高达 80%～90%。主要香气则来自含 3%～5.5%的柠檬醛、α-蒎烯、β-蒎烯等，用于配制饮料、香皂、化妆品及香精。

以粉碎的橙皮为原料，利用水蒸气蒸馏，可以将香精油与水蒸气一起馏出，然后用有机溶剂进行萃取，蒸去溶剂后，即可得到柠檬油。

3.34.3　实验仪器与主要试剂

（1）实验仪器　三口烧瓶（500mL），直形冷凝管，接液管，锥形瓶（50mL、100mL、250mL），漏斗（125mL），

圆形烧瓶（50mL），蒸馏头，温度计（100℃），热浴锅，水蒸气发生器，接液管。

（2）主要试剂　橙皮（新鲜）50g，石油醚，无水硫酸钠等。

3.34.4　实验步骤

3.34.4.1　水蒸气蒸馏

① 安装水蒸气蒸馏装置（见图 2.9）。

② 将 50g 新鲜橙皮剪成碎片后[1]，放入 500mL 三口烧瓶中，加入约 250mL 水。

③ 加热，进行水蒸气蒸馏，控制馏出速度为 2～3 滴/s。收集馏出液约 80mL 时[2]，停止蒸馏。

3.34.4.2　萃取与干燥

① 将馏出液倒入分液漏斗中。

② 用 30mL 石油醚分 3 次萃取。

③ 收集上层液。合并萃取液，

④ 放入 50mL 干燥的锥形瓶中，加入适量无水硫酸钠，振摇至液体澄清透明为止。

3.34.4.3 浓缩

① 将干燥后的萃取液滤入干燥的 50mL 圆形烧瓶中。

② 安装低沸点易燃物蒸馏装置。用水浴加热蒸馏，回收石油醚[3]。

3.34.4.4 产品的收集

当大部分溶剂基本蒸完后，再用水泵减压抽去残余的石油醚[4]。烧瓶中所剩少量黄色油状液体即为柠檬油。

3.34.4.5 测定

通过折射率或旋光度的测定确定产品的纯度。

注释：

[1] 果皮应尽量剪切得碎些，最好直接剪入烧瓶中，以防香精油损失。

[2] 此时馏出液中，可能还有少量油珠存在，但数量已经很少了，可不再继续蒸馏。

[3] 也可用旋转蒸发仪回收石油醚。

[4] 用减压蒸馏可以较彻底脱除石油醚，如用旋转蒸发仪回收，可省去此步。

3.34.5 思考题

① 为什么要采用水蒸气蒸馏的方法提取香精油？

② 用干的橙皮提取柠檬油时，产量会大大降低，这是为什么？

③ 根据本实验，查找相关资料，试提出 1～2 种其他可行的提取植物精油的实验方案。

3.35 黄连中黄连素的提取及鉴定

3.35.1 目的

① 掌握从黄连中提取黄连素的原理和方法。

② 熟悉本实验中的基本操作。

3.35.2 原理

黄连为我国特产药材之一，有很强的抗菌力，对急性结膜炎、口疮、急性细菌性痢疾、急性肠胃炎等均有很好的疗效。黄连中含有多种生物碱，以黄连素（俗称小蘗碱 Berberine）为主要有效成分，随野生和栽培及产地的不同，黄连中黄连素的含量约 4%～10%。含黄连素的植物很多，如黄柏、三颗针、伏牛花、白屈菜、南天竹等均可作为提取黄连素的原料，但以黄连和黄柏中的含量为高。

黄连素是黄色针状晶体，熔点 145℃，微溶于水和乙醇，较易溶于热水和热乙醇中，几乎不溶于乙醚，黄连素存在三种互变异构体，但自然界多以季铵碱的形式存在，其结构式：

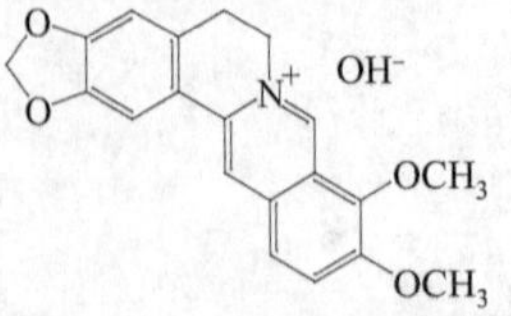

OH^- OCH_3 OCH_3

黄连素的盐酸盐、氢碘酸盐、硫酸盐、硝酸盐均难溶于冷水，易溶于热水，其各种盐的纯化都比较容易。

纯化后的黄连素可通过测熔点、薄层色谱和紫外光谱进行鉴定。

3.35.3 实验仪器和主要试剂

（1）实验仪器　紫外光谱仪，电子天平（0.01g），熔点测定装置，1cm 石英吸收池，不锈钢样品刮刀，圆底烧瓶（250mL），回流冷凝管，水浴锅，吸滤装置，布氏漏斗，烧杯（250、100mL），量筒（25、5mL），广口瓶（50mL），毛细管，铅笔，镊子。

（2）主要试剂　黄连，丙酮（AR），氯仿（AR），甲醇（AR），石灰乳，HCl（12mol/L），醋酸（0.2mol/L），乙醇（95%），氧化铝薄层板。

3.35.4 实验步骤

3.35.4.1 黄连素的提取

① 称取 10g 中药黄连切碎、磨烂，装入滤纸袋中。

② 在圆底烧瓶中加入 100mL 乙醇（95%），安装索氏提取器提取装置（见图 2.6）。

③ 水浴回流 2h，冷却，静置，抽滤。滤渣重复上述操作处理两次，合并三次所得滤液。

④ 在水泵减压下蒸出乙醇（回收），直到出现棕红色糖浆状。

3.35.4.2 黄连素的纯化

① 加入醋酸（0.2mol/L）约 30～40mL 于糖浆状溶液中，加热溶解。

② 抽滤以除去不溶物。

③ 在溶液中滴加 12mol/L 盐酸，至溶液浑浊为止（约需 10mL）。

④ 放置冷却（最好用冰水冷却），即有黄色针状体的黄连素盐酸盐析出（如晶体不好，可用水重结晶一次）。

⑤ 抽滤，结晶用冰水洗涤两次，再用丙酮洗涤一次，加速干燥。

⑥ 将黄连素盐酸盐加热水至刚好溶解，煮沸，用石灰乳调节 pH=8.5～9.8，冷却后滤去杂质。

⑦ 滤液继续冷却到室温以下，即有针状的黄连素析出，抽滤，结晶用冰水洗涤两次，再用丙酮洗涤一次加速干燥。

⑧ 将结晶在 50～60℃下干燥，称量。

3.35.4.3 鉴定

（1）测定熔点　熔点 145℃。

（2）薄层色谱分析　取少量黄连素结晶溶于 2mL 乙醇中（必要时可用水浴加热片刻）。在离薄层板（氧化铝薄层板）一端 1cm 处，用铅笔轻轻划一直线。取管口平整的毛细管插入样品溶液中，于铅笔划线处轻轻点样 1～2 次，晾干或吹干。以 9∶1（体积比）的氯仿和甲醇为展开剂（用硅胶 C 薄层板时，展开剂为 7∶1∶2 的正丁醇-乙醇-水），小心倒入广口瓶中（做展开槽用）。将点好样品的薄层板小心放入瓶内，点样一端在下，浸入展开剂内约 0.5cm。盖好瓶盖，细心观察展开剂前沿上升到离板的上端约 1cm 处时取出，即用铅笔在前沿处划一记号，晾干。计算黄连素的 R_f 值。

3.35.5 思考题

① 黄连素为何种生物碱类的化合物?

② 为何要用石灰乳来调节 pH 值？

3.36 纸色谱分离鉴定氨基酸

3.36.1 目的

① 熟悉纸色谱的基本原理。

② 学会分离氨基酸的方法。

3.36.2 原理

将含有氨基酸的样品点到滤纸上，再将滤纸放到装有少量展开剂的层析缸里。混合物中不同的氨基酸随着溶剂在滤纸上展开，在滤纸上显出不同的斑点，每个斑点表示不同的化合物，这就是纸色谱。如图 2.22 及图 2.23 所示。

吸附在滤纸上的水被称为固定相[1]，溶剂称为流动相。纸色谱是物质在两相的分配过程中滤纸是载体，纸色谱可用来分离混合物。因为物质在固定相和流动相中的分配系数的不同而进行分离。

纸色谱法主要用于分离有色物质，目前也可用于分离无色物质如氨基酸。茚三酮与所有氨基酸反应产生紫色。所以，纸色谱展开后，喷上茚三酮可显出不同的氨基酸，然后测定比移值 R_f。

3.36.3 实验仪器和主要试剂

（1）实验仪器　层析缸，滤纸（5cm×20cm），毛细管（直径 1mm），镊子，烘箱，喷壶，尺子，铅笔。

（2）主要试剂　谷氨酸（0.015mol/L 乙醇溶液），亮氨酸（0.015mol/L 乙醇溶液），谷氨酸和亮氨酸的混合物，茚三酮（0.056mol/L 丙酮溶液），展开剂：正丁醇-乙酸-水（4∶1∶5，体积比）。

3.36.4 实验步骤

① 选一条滤纸（5cm × 20cm）(只能接触一端)，置于一洁净的纸面上，用铅笔在一端从左到右轻轻画一条线（不能用钢笔!），平行从底边 1.5cm 处画三点，点之间距离 1.5cm，表示 A、B 及 A+B。

② 用毛细管点取谷氨酸样品点在标点 A 上，亮氨酸在标点 B 上，混合样在 A+B 上。干燥滤纸几分钟。如果样品点量不足可在同一位置复点一遍，干燥。

③ 按比例向层析缸中倒入展开液，溶液高度约 1.5cm，为了形成饱和蒸气，在展开前要把层析缸密闭几分钟。用铜丝把滤纸条挂在缸中，点样线在下端（点样线不能浸在展开剂里，滤纸边不能接触缸壁）。然后轻轻盖上层析缸，进行展开[2]。

④ 纸色谱展开至少 1h，当展开完成时，应迅速取出滤纸，在溶剂干燥前标出溶剂前沿的位置。

⑤ 在 105℃干燥，然后用茚三酮显色[3]。

⑥ 取回纸色谱滤纸条，放到桌面上，圈出每个斑点测出原点到层析点中心的距离及原点到溶剂前沿的距离。计算 R_f，分析结果。

注释：

[1] 滤纸是纯纤维素，糖类。滤纸表面覆盖着水分子可以被纤维素上的—OH 吸引，水和羟基二者都是极性的，能形成氢键。

[2] 当纸色谱展开时，层析缸不能移动。

[3] 避免茚三酮溶液沾到手或衣服，它会染色很难除掉。

3.36.5　思考题

① 为什么说层析缸里的溶剂的高度不能高于点样线的高度是重要的？

② 为什么用铅笔而不用钢笔标记？

③ 实验中要注意避免滤纸接触层析缸壁，为什么？

3.37　纸上电泳分离鉴定氨基酸

3.37.1　目的

① 学会氨基酸电泳的基本原理。

② 练习用电泳法分离鉴定氨基酸。

3.37.2　原理

带正电荷或负电荷的离子在电场中可以移向与之电性相反的电极，我们把它称为电泳。以滤纸为支撑物的电泳称为纸电泳。

因为我们知道氨基酸的结构取决于溶液的 pH，氨基酸以两性离子存在时的 pH 称为等电点（pI）。在等电点，氨基酸的电荷为零，是电中性的，在电场中不移动。当溶液的 pH 大于等电点时，氨基酸将带负电荷，移向正极。反之，当溶液的 pH 小于等电点时，氨基酸将带正电荷，移向负极。如图 3.5 所示。

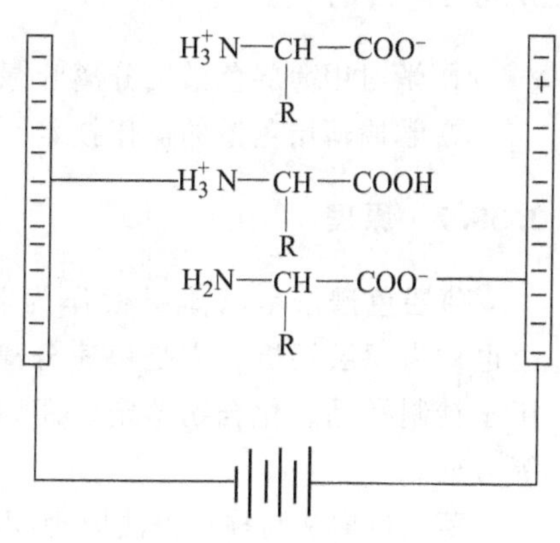

图 3.5　氨基酸的移动方向

按不同的氨基酸离子的不同移动方向，我们可以分离鉴定氨基酸。

3.37.3　实验仪器和主要试剂

（1）实验仪器　DY-2 电泳仪，滤纸（5cm ×30cm），毛细管（直径 1mm），喷壶，烘箱，镊子，铅笔。

（2）主要试剂　谷氨酸（0.015mol/L 乙醇溶液），精氨酸（0.015mol/L 乙醇溶液），谷氨酸和精氨酸的混合物，茚三酮（0.056mol/L 丙酮溶液），巴比妥缓冲溶液（pH=8.9）。

3.37.4　实验步骤

① 取三支毛细管用于点样，毛细管尽量细些。

② 将（5cm×30cm）滤纸条放在一张洁净的纸上，用铅笔在滤纸中间画一条线，标记三个点 A、B 和 C，间距为 1.5cm。小心不要用手指接触滤纸。

③ 用毛细管点样，使其干燥几分钟。再复点 1～2 次，再使其干燥。样品 A 是谷氨酸，

B是精氨酸，C是混合样。

④ 将点好的滤纸放在电泳仪的架子上，滤纸的两端滴上缓冲液，用缓冲液润湿滤纸点样线1cm外的部分直到溶液渗透点样线，然后盖上电泳仪。

⑤ 打开电源，调节电压在220～280V之间，约30min后，关闭电源。

⑥ 电泳完成时，用镊子取出滤纸挂起烘干，然后用茚三酮显色，在105℃再干燥直至呈现斑点，圈出斑点。

⑦ 分析实验结果。

3.37.5 思考题

① 在本实验中，谷氨酸移向哪一极？为什么？

② 纸色谱和纸电泳在原理上有什么不同？

③ 脯氨酸的pI=6.4，天冬酰胺的pI=5.4，在pH=8.9的缓冲溶液中，将以何种荷电状态存在？

3.38 薄层色谱法分离叶绿素

3.38.1 目的

① 学习用薄层色谱法分离叶绿素。

② 掌握薄层色谱的操作技术。

3.38.2 原理

薄层色谱法是以薄层板作为载体，让样品溶液在薄层板上展开而达到分离的目的，故也称为薄层层析。它是快速分离和定性分析少量物质的一种广泛使用的实验技术，可用于精制样品、化合物鉴定、跟踪反应进程和柱色谱的先导（即为柱色谱摸索最佳条件）等方面。

薄层色谱法与柱色谱法原理相同且也是固-液吸附色谱。在薄层色谱里固定相被以薄层（约250μm）形式涂在玻璃板上。分离的物质被点在薄板上，然后把薄板放入盛有足够的展开剂的层析缸中（不能没过点样线），混合物中的组分则以不同的速率随着溶剂的展开。

计算各成分的R_f值，同时，为了保留薄板上的信息，将展开板上溶剂前沿、点样点、各斑点按比例画在记录本上。

3.38.3 实验仪器和主要试剂

（1）实验仪器　研钵，分液漏斗，层析缸。

（2）主要试剂　硅胶板，石油醚，乙醚，新鲜菠菜，无水硫酸钠，乙醇，氯仿。

3.38.4 实验步骤

在研钵中，放几片新鲜菠菜叶和几毫升2∶1的石油醚（b.p.30～60℃）和乙醚（体积比）溶液，仔细研磨。用吸管将液体转入小分液漏斗，加等体积水，旋摇，如振摇液体可能乳化。静置分层，分出、弃去下面的水层。有机层再用水洗涤两次，水相弃掉。水洗可除去乙醇及从叶子里提取的水溶性物质。将有机层倒入小锥形瓶，加2g无水硫酸钠干燥。放置几分钟后倾出溶液，如颜色不深，将溶液浓缩。然后取 10cm ×2cm 硅胶板，用毛细管将带色溶液点在硅胶板上，样品点距末端约 1.5cm。点样时，应避免样品点直径扩散超过 2mm。将硅胶板晾干，用氯仿作展开剂，放入层析缸中展开。

展开后的硅胶板上有时可能会有八个有色斑点，按 R_f 递减，这些斑点已被确证：胡萝卜素（两个点，橙色）、叶绿素 a（蓝绿）、叶黄素（四个点，黄色）和叶绿素 b（绿色）。

计算展开版上所有斑点的 R_f 值，同时，为了保留板上信息，将展开版上溶剂前沿、点样点和各斑点按比例画在记录本上。

注释：

使用石油醚萃取叶绿素时，附近不能有明火，因其非常易燃。

3.38.5　思考题

① 进行 TLC 试验时，为什么样品点不能浸入层析缸的溶剂里？

② 解释为什么薄层板展开时不能让板上的溶剂蒸发？

参考文献

[1] 陆涛，陈继俊．有机化学实验与指导．第2版．北京：中国医药科技出版社，2009.

[2] 李英俊，孙淑琴．半微量有机化学实验．第2版．北京：化学工业出版社，2009.

[3] 袁华等．有机化学实验．北京：化学工业出版社，2008.

[4] 卢艳花．中草药有效成分的提取分离实例．北京：化学工业出版社，2007.

[5] 薛思佳，Larry Olson. 有机化学实验（双语教材）．北京：科学出版社，2005.

[6] Department of chemistry Xi'an jiaotong University. Experimental Organic chemistry. Xi'an：Xi'an jiaotong University，2002.

[7] 周志高，蒋鹏举．有机化学实验．北京：化学工业出版社，2004.

[8] 李霁良．微型半微型有机化学实验．北京：高等教育出版社，2003.

[9] 杨善中，柴多里等．有机化学实验．合肥：合肥工业大学出版社，2002.

[10] 马全红，路春娥等．大学化学实验．南京：南京大学出版社，2002.

[11] 李兆陇，阴金香，林大舒．有机化学实验．北京：清华大学出版社，2001.

[12] 曾昭琼．有机化学实验．第2版．北京：高等教育出版社，2000.

[13] Harwood L M，et al. Experimental Organic Chemistry，2nd ed. Oxford：Blackwell Science Ltd（add：OsneyMead，Oxford OX2 0EL），1999.

[14] 周科衍，高占先．有机化学实验．第3版．北京：高等教育出版社，1996.

[15] 谷亨杰．有机化学实验．北京：高等教育出版社，1991.

[16] 北京大学化学系有机化学教研室．有机化学实验．北京：北京大学出版社，1990.

[17] Charles F，Wilcox Jr. Experimental Organic Chemistry. MacMillan Publishing Company，Adivision of MacMillan Inc，1998.

[18] 帕维亚 D L，兰普曼 G M，小克里兹 G S. 现代有机化学实验技术导论．丁新腾译．北京：科学出版社，1985.

[19] 费歇力 L P，威廉森 L. 有机实验．左育民等译．北京：高等教育出版社，1980.

[20] 印永嘉．大学化学手册．山东：山东科学技术出版社，1985.